Rise and Fall of The Shadow Government

Return of the Mother Plane

Revised Edition 2014

Copyright©2010

(Formerly named "The Classical Gold Standard: Rise and Fall of Caucasian Civilization")

The title of this book was changed based upon the final 4 "Time and What Must Be Done" series presented by the Honorable Minister Louis Farrakhan

Written By Rasheed L Muhammad

I dedicate this book to Bro. Ron and Sister Constance Muhammad

of

Muhammad Mosque 32

Contents

Introduction

In this book, *"Rise and Fall Of The Shadow Government: Return Of The Mother Plane"*, what you will read is the history about how Signore Families (i.e. Banking Families living independent of legal fictional laws that govern the common people) actually represent the shadow government.

Although today the Rothschild's may be known as the most famous Signore family of our time, there were several before them. Take for instance the Medici family, the Hapsburg family and the Capetian family. These Caucasian or European Signore Families have placed Popes into office, financed private armies, printed and lent their form of money and persuaded kings and queens to govern lands according their truth even if it were against divine law and sound scripture. These families and other notable omissions of Europeans or Caucasian Signore families; namely, Merovingians, Tudors, Romanovs, Stuarts, House of Orange, Berenguer family, Kennedy family and Vanderbilt families remain linked through marriage. This method has assured the consolidation of their wealth through a monopoly of holding real Gold in their private vaults stashed around the world.

At the gates of Jerusalem, Jesus of 2000 years ago saw such an anti-God money-system developing between the money- changers and Roman authorities. When they began and allowed *Coinage clipping (mixing* copper with *Gold and Silver)*, thus corrupting and devaluing the median of exchange, Jesus simply referred to that shadow government as "the synagogue of Satan."

"I will cause some belonging to Satan's synagogue who say that they themselves are Jews, and are not, but are liars--I will make them come and fall at your feet and know for certain that I have loved you." (Revelations 3:9 Weymouth New Testament)

Unrelenting to achieve world dominance, the last and most powerful government to fall under its control under the Synagogue of Satan was/is what we call the United States of America. You ask in what why? Allowing for a Signore Family Central Bank to print all U.S. currency and to dictate monetary policy (rise and fall of usurious interest rates constructed upon "debt" credit).

In an online article entitled THE SECRET SHADOW GOVERNMENT: A STRUCTURAL ANALYSIS written by Dr. Richard J Boylan Ph.D, he says, "The secret shadow government is the large organizational network which operates alongside the officially elected and appointed government of the United States of America. Just as with the official government, the secret government has functional branches."

In this book *"Rise and Fall of the Shadow Government: Return of the Mother Plane"* you shall read by what means shall this shadow government be brought down to its knees. The scripture warned them thousands of years ago that God is a Man of War. It reads, *"The Lord is a man of war,"* in Exodus 15:3.

Therefore, the final chapter of this book *"Rise and Fall of the Shadow Government: Return of the Mother Plane"* discloses the weapon of choice (called the Mother Plane) created by God—Supreme Being and His Army of Scientists whom the scripture call Angels that will strike from the sky against the weapons made by the military scientists employed by the shadow government whom God calls Satan.

Rasheed L. Muhammad
February 16, 2010
Revised February 15, 2014

Chapter ONE

Secret Gold Addiction

Gold has been the foundation of monetary systems for centuries. During the Classic period of Greek and Roman rule in the western world, gold and silver both flowed to India for spices and to China for silk. When money was invented, its name was Gold. At the height of the Empire from 98-160 A.D., Roman gold and silver coins reigned from Britain to North Africa and Egypt.

Members of Caucasian race began pillaging our earths Gold from Guinea in 1663 to Brazil in 1702, from California in 1848 to Australia in 1851, from South Africa in 1886 to Western Australia in 1893, and finally in Canada from 1897 to 1899. As for the natives of these lands, their Gold was "strong armed". Consequently, poverty and debt is what they suffer today. Many of these countries now sell their birthright (land) just to receive food-aid and aid from the handouts of Shadow Government networks and agencies.

In 1914, the British Empire or *shadow government* opened the door to a new world paper currency with the intention to replace the ancient Gold Standard invented by the ancient Black nations of the earth. So after 1914 Westernized governments began borrowing paper money which generated monetary inflation and insurmountable debt for all subjected the rule of the West.

"The gold specie standard ended in the United Kingdom and the rest of the British Empire at the outbreak of World War I. Treasury notes replaced the circulation of gold sovereigns and gold half sovereigns. Legally, the gold specie standard was not repealed. The end of the gold standard was successfully effected by the Bank of England through appeals to patriotism urging citizens not to redeem paper money for gold specie." [1]

This scheme is what has enabled Signore families (Central banking families) to suppress, at will, all governments we know today (under the International Monetary Fund system) by misusing what is called "Gold debasement" (watering down true value of Gold with disproportionate printed paper money). *See pg. 123 Appendix 1: List of IMF Members.*

In past times Gold, silver or copper coins generated public confidence for knowing that its Government held Gold and other valuable metals in safe keeping. The question is: Does America hold enough tonnes of the precious yellow metal to justify her total output of M3 (money supply) ratio for its Gold holdings to move her Gold debasement (watering down the true value of Gold). That is the question!

In economics, money supply or money stock is designated as M3—the total amount of money available to an economy at a particular point in time. As of late, M3 is no longer published or revealed to the public by

---

[1] http://en.wikipedia.org/wiki/Gold_standard#Impact_of_World_War_I

US government officials due to massive government fraud. Therefore, the plausibility of fraud sits at the foundation of this world because its M3 output has debased (watered down) the value of its Gold holdings thereby destroying the wealth of society and our true purchasing power.

For instance: The value of US Gold reserves is approximately 1.2% of the value of M3 as of June 2, 2005. M3, at that time, (some say was $8.73 trillion, or to put it the long way, "one cent of Gold's value is used as collateral per 83 U.S. dollars. So, roughly, you can say that each U.S. $1 bill is 'backed' with about $0.012 cents worth of Gold - a ratio of 83 to 1.1 Is this a recipe for disaster? Of course it is! These figures only relate to 2005, 2013 figures are as follows:

> "The U.S. money supply just keeps growing. In August 2013, M1 was $2.55 trillion and M2 was $10.77 trillion. For M1, slightly less than half ($1.2 trillion) was cash and travelers checks, while the rest was checking accounts. For M2, $7 trillion was in savings accounts."[2]

This problem is so horrible; no government official is allowed to speak about it in detail. Though, the above quote is similar to saying the U.S government is employing far less than 1 cent of Gold's value as collateral per 83 plus U.S. dollars.

How did paper become more valuable than Gold? Answer: When the people of the planet were

---

[2] http://useconomy.about.com/od/inflationfaq/f/Causes-Of-Inflation.htm

conditioned to accept paper money as more valuable than Gold. For this, 188 IMF nations, including the United States, live in debt up to our ears to a most diabolical shadow government of Signore families who care no more about us as they did about Jesus when they discovered he was a black man entombed in glass at the Mosque of Omar in old Jerusalem.[3]

Of course Gold debasement is not a new practice among the criminal mind within the circles of the shadow government *shot callers*. Gold debasement was implemented by kings of England centuries ago but with a slight variation when the Crown's policy of coinage debasement earned them great profits. Yes under Henry VIII and Edward VI **Coinage debasement occurred when their way of ruling and cheating the people began mixing a significant degree of a base metal such as copper with Gold or silver coinage content to dilute Gold coin value. By reducing the value of Gold or silver content relative to its real face value, government can extract usable revenue from domestic money stocks.** This criminal stratagem is called debasement because each coin is worth less in terms of its precious metal content.[4]

In 2014 a significant degree of Central bank paper-money rather than base metals are being substitued to devalue Gold reserves of all nations including America.

---

[3] The Book Of God: An Encylopdia of Proof That The Black Man Is God, pges. 230 – 249 by True Islam (aka Dr. Westley Muhammad)
[4] http://en.wikipedia.org/wiki/Great_Debasement

To what degree you ask: Unto the fall of the Caucasians civilization because America's Gold debasement scheme has functioned under bankruptcy protection since 1933. America is simply 1 of the 188 member nations under the International Monetary Fund (IMF). This organization and its member countries hold Gold and share its yellow metal to pool among one another as if they are toying with Africa's ancient Susu economic system.

In the book, "Susu and Susunomics: The Theory and Practice of Pan-African Economic, Racial and Cultural Self-Preservation by Paul Barton, he espouses: *"Susu economics is a collective of people pooling their resources together for the benefit of those within pool. Each participant in the pool is on a time sensitive commitment to a turn to employ funds raised by members of the Susu to purchase or acquire goods that will produce profits. Essentially, 'Susu is a trust and commitment of collective capitalism.'"*

The difference; however, between the ancient "Susu' system and the IMF system is that IMF rules in a criminal manner to suppress the general populations spending power by stealing the people wealth and resources on the cheap. By his I mean Gold reserves are almost exclusively used in the settlement of international transactions via the IMF. If you are not a member, you cannot gain access to its pool of Gold. Their most suppressive mathematical formula that Governs access to the Gold pool is as follows:

Interest = Z p

where

$$\frac{d}{360}$$

Z = amount of metal (oz.) being loaned
p = metal lease rate
d = number of days in accrual period

Example: Borrow 1,000 oz. of gold for 2t a lease rate of 1% p.a. The interest you owe is 20 oz. (=1,000oz. x 1% x 720/360)

Basically the Central Bankers and their Gold Bullion houses can borrow/lease gold at discounted rates only to return it after making huge profits. With that money or profits, they buy real resources next to nothing. Another way to explain the above formula is to say Central banking families like to hold gold reserves to earn income on their gold holdings. So they loan out a large proportion of their gold reserves in the gold loan market. The market is called a forward market, as it is a synthetic (insincere: not genuine) commodity lending market. In the case of gold, it is an actual commodity loan market. In consequence, interest on gold (and other metals) loans is calculated and payable in metal. The wickedly wise shadow government understands this business and greatly manipulates it through their networks and organizations to ruin an honest money supply. In other words, while they use Gold among themselves to conduct big government business cheaply, they print paper-money for non-Signore Families to use at exurbanite inflationary cost.

"Inflation is the devastating condition when prices just keep going up, eating away at your standard of living. There are three main causes: demand-pull, cost-push, and monetary expansion." [5]

Another example to explain the above occurred in 1980 when Central Banks began to lease or lend Gold at 1% to Bullion banks who then employed the Gold as an investment. Sometimes these investments earned 6% returns on government Bonds. When this happens, civilization in general losses as our intrinsic wealth is leased to a few who practically steal more of the people's wealth on the cheap turning borrowed Gold into a profit. You ask how they get away with it. Answer: through the army of law makers (legislators) trained and educated in Europe and America's prestigious international universities and private colleges.

I say again, Gold debasement is fixed to the advantage mainly for Central bankers and their insider networks of bullion banks—the wealthiest of the wealthy that borrow a "governments" Gold horde as an asset. Then they use those assets to purchase high value resources with man-made paper money before returning the Gold back to Central bank vaults with

---

[5] http://useconomy.about.com/od/monetarypolicy/f/money_supply.htm

interest in the form of a few ounces of gold. The crime is the more paper money printed the less value the Gold is fixed and ultimately the less value the paper money. So by the time the paper money reaches you and I (consumers), it has lost too much value or buying power. So goods are not going up in price, it's more like the dollar is losing its value

London Gold Fix

As it were, the London Fix began in 1919 when the five representatives met at Rothschild's office. During those years, Gold was fixed at four pounds, 18 shillings and nine pence equal to about $7.50 U.S. dollars in today's terms.

> *"The London gold fixing or gold fix is the procedure by which the price of gold is determined twice each business day on the London market by the five members of the London Gold Market Fixing Ltd. It is designed to fix a price for settling contracts between members of the London bullion market, but informally the gold fixing provides a recognized rate that is used as a benchmark for pricing the majority of gold products and derivatives throughout the world's markets.*

> *"The gold fix is conducted in United States dollars (US$), Pound sterling (GBP), and the euro (€) daily at 10.30am and 3pm, London time, via a dedicated telephone conference facility. The current five participants in the fixing, who must be members of the London Bullion Market Association, are:*

> *"Scotia-Mocatta — successor to Mocatta & Goldsmid and part of Bank of Nova Scotia Barclays Capital — Replaced N M Rothschild & Sons when they abdicated Deutsche Bank — Owner of Sharps Pixley, itself the*

merger of Sharps Wilkins with Pixley & Abell HSBC — Owner of Samuel Montagu & Co. Société Générale — Replaced Johnson Matthey and CSFB as fifth seat."[6]

Understanding this history manifests how Gold insider trading profits these Signore families though their corporations. *See Appendix 1, London Gold Fix, pg. 132*

Some gold experts believe were it not the crafty manipulations of the Gold fix, Gold bullion's value would be around $25,000 per/oz.

> *"Will such increased demand for gold, and price inflation of gold, come? The United States is already pumping money at very strong levels and the euro crisis may result in printing by the European Central Bank, which will magnify price inflation on a global scale. The printing in the United States will likely be enough to push gold much higher...and likely result in a lot of new buyers beyond just the current gold bugs – stepped up buying by central banks, buying by the very wealthy and buying by the average American. If global inflation occurs, $25,000 per troy ounce may be a conservative estimate as to where the price of gold eventually goes. "[7]*

If this were the case and Gold' value is actually fixed in ratio to M3 Central bank paper money, all IMF member countries would cease commerce operations and crash. Hence, accounting fraud is a necessary evil for Central bankers and the governments under their

---

[6] http://en.wikipedia.org/wiki/Gold_fixing
[7] http://www.munknee.com/gold-could-reach-25000ozt-heres-why/

command and control. By maintaining authority over the Gold supply of entire governments or countries, and placing certain elite-well educated and aristocratic sycophants in decision making seats within those governments, legal fiction and financial illusion continues.

Most Signore families' main goal is to usher in a one world government with one currency. They believe we all should say, "Who else do you trust to run our worlds' money supply, China, Russia, Arabia...?" And if total resistance to their goal is threatened, they're considering bankrupting the entire global market.

You may not know it, but global bankruptcy happened 650 years ago in Europe, and it might be repeated if global conformity is resisted over the next 3 to 4 years. At the moment, China and the Muslims world is resisting while the Holy See is considering Islamic banking because he knows Western capitalism is ruined due to greed and fraud. So stay tuned!

Chapter TWO

Rise Of The Gold Standard

Gold was made into coins to enhance its usability as a monetary unit. Roughly speaking, in 2,700 BC, the first Gold coins were issued by Egyptian pharaohs. Then around 560 BC King Croesus of Lydia (modern-day western Turkey) produced large scale Gold coinage for the monetary purposes. He was once the richest man ever.

Did aboriginal nations use Gold for worship? I say no more than people of modernity worship Gold around their necks, wrist and fingers. Today around 70% of Gold demand is jewelry, 11% is industrial (dental, electronics) and 13% is investment (institutional and individual, bars & coins).

Since Europe's elite could not rely on additional Gold supplies from West Asia, the supply of Gold expanded only through deflation, trade, pillage and fool's gold. As a result, their obsession for Gold led to conquering Africa, North America, South America and other aboriginal lands for Gold and other raw materials. The discovery of America in the 15th century led to a further obsession for Gold. Spain's plunder of treasures from the New World raised Europe's supply of Gold five-fold in the 16th century.[8]

Other Gold rushes in America, Australia and South

---

[8] www.romanbritain.freeserve.co.uk/1696recoinage.htm

Africa took place during the 18th and 19th century.

Prior to Gold coinage use as a money unit, Gold was weighed and checked for purity when settling trades. When coinage became the norm, debauched Gold coins or clipped Gold coins were always something business traders had to be aware of—the mixture of other base metals mixed with Gold thereby lessening its value. In 1696, the Great Recoinage in England introduced a technology that automated the production of coins, and put an end to clipping.[9]

While Gold coins and Gold bullion usage continued to dominate the monetary system of medieval Europe and the world for that matter, it was not until the 18th century that paper money began to debase (clip) the value of Gold. Since there is no unlimited amount of Gold, eventually a Gold standard was agreed upon by Caucasoid nation states. Between 1696 and 1812, Europe was inching up toward the development and formalization of its Gold standard since they had introduced paper money into their system of exchange. What they discovered is the ratio of Gold to paper money must maintain a balance to maintain an honest money supply.

"In 1797, due to too much credit being created with paper money, the Restriction Bill in England suspended the conversion of notes into Gold... A Gold standard was needed to instill the necessary controls on

[9] www.romanbritain.freeserve.co.uk/1696recoinage.htm

money. By 1821, England became the first country to officially adopt a Gold standard...from 1871 to 1914; the Gold standard was at its pinnacle. During this period near-ideal political conditions existed in the Caucasian's civilization.

Governments worked very well together to make the system work, but this all changed forever with the outbreak of the Great War in 1914. By 1914, Europe's Gold standard was ruined because of tribal warfare, greed and over expansion of what they called "global economics". Fact is, there was not enough total Gold supply to back bank note gold and silver certificates. Yet Europe's internecine warfare benefited Central banks so they continued to issue loans to all sides of the warring factions. More loans equaled more debt for governments; more debt equaled more interest payments to Central banks. Thus by issuing more bank notes caused further Gold debasement.

Further Gold debasement meant inexpensive Gold, which was purchased by Central banks and private financial firms. Consequently Europe's paper printing Central bank networks organized to establish and to maintain a monopoly over all real wealth by undermining the classical Gold standard. What is a classical Gold Standard?

A classical Gold standard is a monetary system in which bank notes are freely convertible into a fixed amount of Gold. When this system is not in place, and people cannot convert a bank note gold certificate into a fixed amount of Gold, it means your government has no control over its wealth and purchase power of its citizenry.

For example, after 1914, Europe's Central banking cartel pressed for a one-world-government monetary system. This meant all nations would become debtors under white-authority and they, the private owned bankers of Europe, would become creditors because religious leaders fell under the seven deadly sins:

1 Lust (Latin, luxuria)
2 Gluttony (Latin, gula)
3 Avarice (Latin, avaritia)
4 Sloth (Latin, acedia)
5 Wrath (Latin, ira)
6 Envy (Latin, invidia)
7 Pride (Latin, superbia)

Fall of the Gold Standard

In 1914 European nations fought World War I (WWI) to see who would control the resources of the aboriginal nations; namely, Africa. The curse of this police action unraveled Europe's Classical Gold Standard. I say this because Central bankers are not the wisest of the wise. They were merely fulfilling a page of the divine script until the end of a certain TIME one may call debt financing.

It appears all too often, Caucasoid arrogance prevents their elite from remembering why God's wisdom was rehearsed by ancient nations before them and nothing new today is being practiced from times past. Greed and warfare and fraud have always terminated a government's progress and existence.

After Western Europe evolved out of the dark ages of the 13th century and devised how to employ its Classical Gold Standard into a monetary system that created near-ideal political conditions from 1871 to 1914, it all came crashing down. After WWI political alliances changed, international indebtedness occurred and government financial deterioration was evident.

Only a few knew this is what the Central bankers (Signore families of Europe) desired to achieve. That is subject the populous under their rules and regulations otherwise known as legal fiction. Banking families were able force sovereign governments to leave the Gold Standard in exchange for their fiat money. Thereafter, a great fraud was initiated to extend Western Europe's central administrative apparatus into the technological age to conquer, by debt, nations and people outside of Europe as well as inside Europe.

By this I mean, after WWI the Bank of England suspended its gold standard. This act ultimately led to their civilizations need for something more flexible on which to base its global economy. What was more flexible than Gold? Of course, privatized paper-printed bank notes, which was another redo of the same old tired concept used by ancient China's Song and Ming Dynasty's mainly from A.D. 960 - 1644. But, paper money did not endure then and it will not endure under the Western World's current fiat money world order.

Nevertheless, as Europe's Central bank continued with its Gold debasement fiddle to push for the growth of its fiat global economy, the British pound sterling and U.S. dollar became the global reserve currencies after 1914. Following this act of financial fraud, naïve smaller countries began holding more of these currencies instead of Gold, resulting in an accentuated consolidation of Gold into the hands of Europe's most powerful governments.[10] Then fifteen years later, in 1929, came the panic and economic crash of the United States of America. What happened during the years before and after the crash? Well most countries in Europe raised their interest rates hoping their national and foreign investors would keep deposits in banks. But investors wanted to withdraw their money for its value in Gold since the Gold standard had not yet been completely eradicated.

And since Europe's Central bank of England needed to prevent a run on its Gold supply from leaving its vaults in an un-orderly fashion the Gold standard in England was suspended in 1931, leaving only the U.S. and France with large Gold reserves. But finally in 1933, America became the final victim of England's international bank plan to force the U.S. off the Gold standard for access to fait or paper money.

---

[10] www.investopedia.com/articles/05/030705.asp

To bring this about, the U.S. was made to declare bankruptcy thus forcing all Americans to convert their gold coins, bullion and certificates into U.S. Federal Reserve Notes. This act was designed to avert the outflow of gold during the Great Depression, so we are told.

"...in 1934, the U.S. government revalued Gold from $20.67/oz to $35.00/oz, raising the amount of paper money it took to buy one ounce, to help improve its economy. As other nations could convert their existing Gold holdings into more U.S dollars, a dramatic devaluation of the dollar instantly took place. This higher price for Gold increased the conversion of Gold into U.S. dollars effectively allowing the U.S. to corner the Gold market. Gold production soared so that by 1939 there was enough in the world to replace all global currency in circulation.

"As World War II was coming to an end, the leading western powers met to put together the Bretton Woods Agreement, which would be the framework for the global currency markets until 1971. At the end of WWII, the U.S. had 75% of the world's monetary Gold, and the dollar was the only currency still backed directly by Gold.

"But as the world rebuilt itself after WWII, the U.S. saw its Gold reserves steadily drop as money flowed out to help war-torn nations as well as to pay for its own high demand for imports. The high inflationary environment of the late 1960s sucked out the last bit of air from the Gold standard.

"In 1968, a Gold pool (which dominated Gold supply) which included the U.S and a number of European nations stopped selling Gold on the London market, allowing the market to *freely determine the price of Gold. From 1968 to 1971, only central banks could trade with the U.S. at $35/oz. Finally, in 1971, even this bit of Gold convertibility died."*[11]

In later chapters of this book, I will delve more into the year 1933 and the fall of the gold standard. For now the question is: Do Central banks hold about 18% of all gold ever mined? And if they do, who holds the other 82%. Answer: it doesn't matter at the moment because Central banks print all of the M3 (paper money or fait money) to reduce Gold's real value. Categorically their Gold and paper money monopoly is fixed straight out of London and enforced by the U.S. Federal Government to maintain a false perception of white male masculinity and dominance over the aboriginal nations of the planet.

---

[11] www.investopedia.com/articles/05/030705.asp

Chapter THREE

Arabia's Ancient Gold Mine

Howstuffworks.com explains Gold has a specific gravity of 19.3, meaning that it is 19.3 times heavier than water. So gold weighs 19.3 kilograms per liter. A liter is a cube that measures 10 centimeters (about 4 inches) on a side. There are 32.15 troy ounces in a kilogram. Therefore, the world produces a cube of gold that is about 4.3 meters (about 14 feet) on each side every year. In other words, all of the gold produced worldwide in one year could just about fit in the average person's living room! Some gold bugs estimate 158,000 tonnes of Gold has been mined since the beginning of time.

The earliest preserved geologic map was made in 1150 BCE to show the location of Gold deposits in Eastern Egypt; it is known as the Turin papyrus. The Greek name for Aswan, Syene; is the type locality for the igneous rock syenite. The Romans followed this tradition and had many quarries especially in the northern part of the Eastern Desert of Egypt where porphyry and granite were mined and shaped for shipment. The most productive mine in Saudi Arabia, Mahd adh Dhahab ("Cradle of Gold"), has been periodically exploited for its mineral wealth for hundreds or even thousands of years and is reputed to be the original source of King

Solomon's Gold.[12]

The Holy Land is the ancient Arabian Peninsula—the once land of Gold according to Genesis 2:11-14, *"The name of the first is Pishon: that is it which compasseth the whole land of Havilah, where there is Gold; and the Gold of that land is good: there is bdellium and the onyx stone. And the name of the second river is Gihon: the same is it that compasseth the whole land of Cush."* Of course, Cush (Ethiopia) also means land of black people. So supporting the Biblical map below is evidence of Havilah, in Arabia where the river Pishon enters as shown in illustration I below.

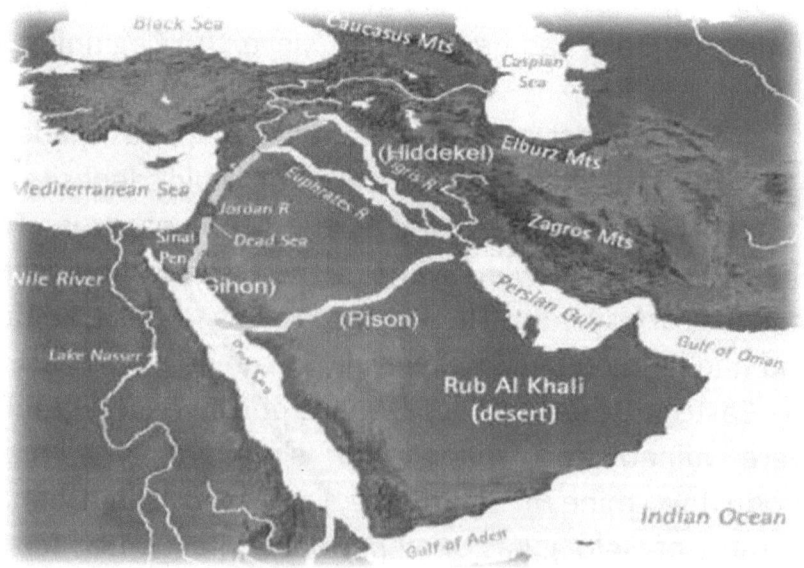

Gold was mined in the Holy Land 7,000 years ago

---

[12] http://en.wikipedia.org/wiki/Mahd_adh_Dhahab

and maybe even beyond. Gold deposits in Egypt and northern Sudan were found and exploited by Egyptians, but new Gold discoveries have been found in Sudan, Eritrea, and in Black Saudi Arabia as far back as 5,000 years ago some scholars claim. By ancient accounts, Black people (Cush) of the earth were aware of Gold long before Western Caucasians were aware that the earth was round, let alone the value of Gold. This entire area including Arabia was once known as the holy land east, west, north and south.

> *"The Gold of Arabia was famed in antiquity." Hastings identified Havilah as "the 'region of northern Arabia, which extended westward towards the frontier of Egypt."*[13]

The Honorable Elijah Muhammad taught the root of civilization began in Arabia, whose holy precinct is the city of Mecca. Mecca is a more ancient "first" religious free enterprise zone established by the original Black nation even before prophet Abraham and Ishmael were born, Quran 3:95. And certainly before Vatican City was established in 1929.

What is a free zone? A "Free Zone" is an area, geographically limited, where the local and federal governments "back off" to some extent to permitting many incentives in such a zone to encourage investors; tax credits, relaxation of licensing restrictions, streamlining of permits processes, variances to zoning laws and a general reduction of restrictive regulations. In

---

[13] http://historicalgenesis.com/synopses.aspx?chapter=03

other words, less government allows wealth to flow under righteous law and order.

Holy Land Once Governed by Black People

Arabia—the Holy Land and/or the entire area of Paradise (north, east, south and west)was ruled by original Asiatic Black men and women of the earth. There were no red Arabs (what is now called white skin Arabs) on the Arabian Peninsula, only Black and Brown. The white skin Arabs or red Arabs and/or Semitic people emerged formed in two different time periods. First around 6,600 years ago white skin people and then strongly Semitic people around 1,700 BC.

Prophet Mohammed of Arabia was born 570 years after Jesus (who was a jet black man). This information is important for our discussion because Mohammed of Arabia was a descendent of Black Arabs. His paternal grandfather, 'Abd al-Muttalib, sired ten black sons all pure Arabs.

The Syrian scholar and historian al-Dhahabi (d. 1348) also reported that 'Abd Allah b. 'Abbas, Muhammad's first cousin, and his son, 'Ali b. 'Abd Allah, were "verydark-skinned." 'Ali b. Abu Talib, first cousin of the Prophet and future fourth caliph, is described by al-Suyuti and others as "husky, bald...pot-bellied, large-bearded...and jet-black (shaded al-udma)." 'Ali's son, Abu Ja'far Muhammad, according to Ibn Sa'd (d. 845), described 'Ali thusly: "He was a black-skinned man with big, heavy eyes, pot-bellied, bald, and kind of short."[14] So rather we are talking about the root of civilization (Mecca) or Mahd adh Dhahab (literally meaning "cradle of Gold"), the Peninsula of Arabia once upon a time was not only governed by Black men and women, it also yield the richest Gold mines of ancient world civilizations.

Furthermore, it [Mahd adh Dhahab] is believed to be the fabled "Ophir" of the Bible, the source of King Solomon's Gold. (Ophir was another one of Joktan's sons; (Gen. 10:29.) The Gold of Ophir is referred to in the following passages: 1 Kings 9:28, 10:11, 22:48; 1 Chron. 29:4; 2 Chron. 8:18, 9:10; Job 22:24; Ps. 45:9; and Isa. 13:12. Based on the number of ancient mine tailings (refuse left over after the ore is treated), geologists have estimated that the Mahd adh Dhahab mine produced more than 950,000 ounces (about 30 metric tons) of Gold in antiquity and they believe it was mined during the reign of King Solomon (961-922 B.C.) and Gold may have been mined at Mahd adh Dhahab much earlier than during Solomon's time--even as early as the patriarchal period as evidenced by the [Gold jewelry found in the] *Royal Tombs of* Ur."[15]

Notice Mahd adh Dhahab is located N.E of Mecca and S.E. of Medina no more than 150 to 200 miles in either direction. The question is: Who is the original man?

---

[14] http://blackarabia.blogspot.com/2011/09/bilal-b-ribah-not-first-black-muslim.html
[15] http://en.wikipedia.org/wiki/Ophir

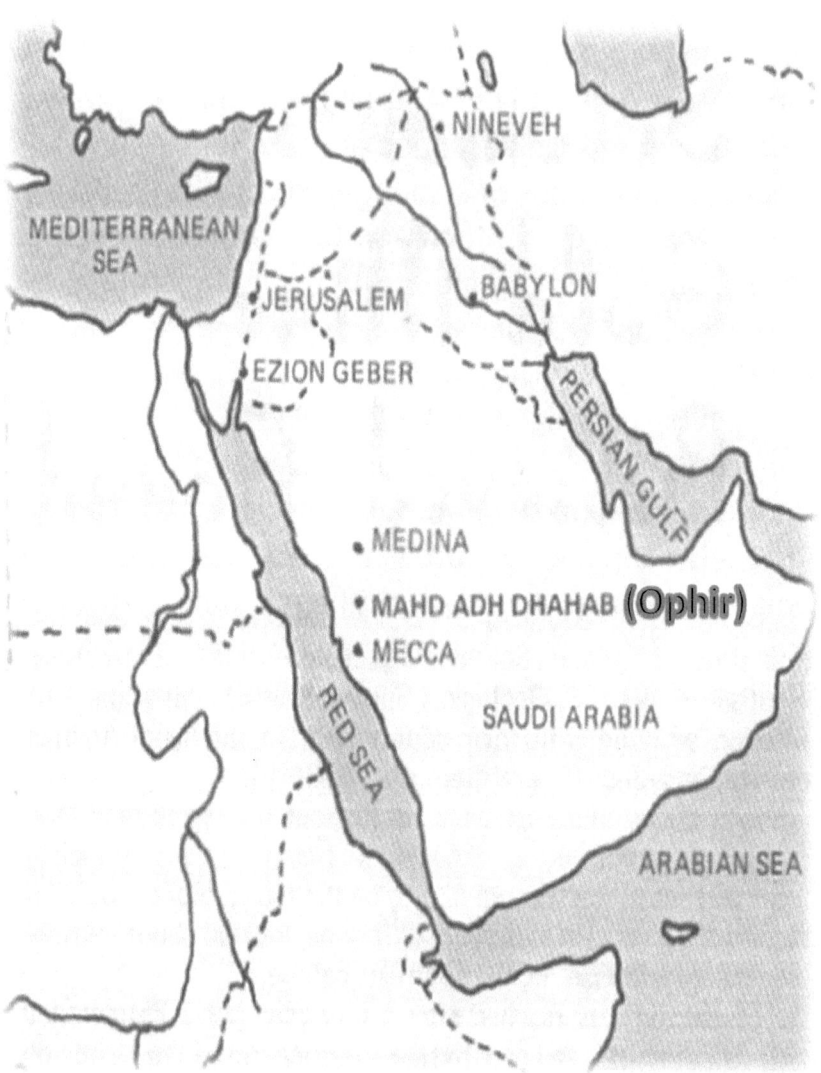

Chapter FOUR

Golden Age Of Mali

Mali once controlled trans-Saharan trade in Gold, salt, slaves, and other precious commodities of West Africa. Ghana was the earliest of these empires. In 1307 (during which time Europe was in its dark ages) Musa I of Mali was the tenth mansa, meaning "king of kings", of the Malian Empire. He reigned for over twenty years. At the time of Mansa Musa's rise to the throne till his demise, he perhaps was the wealthiest ruler of his day. This African Islamic ruler is credited with bringing about the Golden Age of Mali. At the time of Mansa Musa's rise to the throne, the Malian Empire in map below consisted of territory formerly belonging to the Ghana Empire of West Africa.

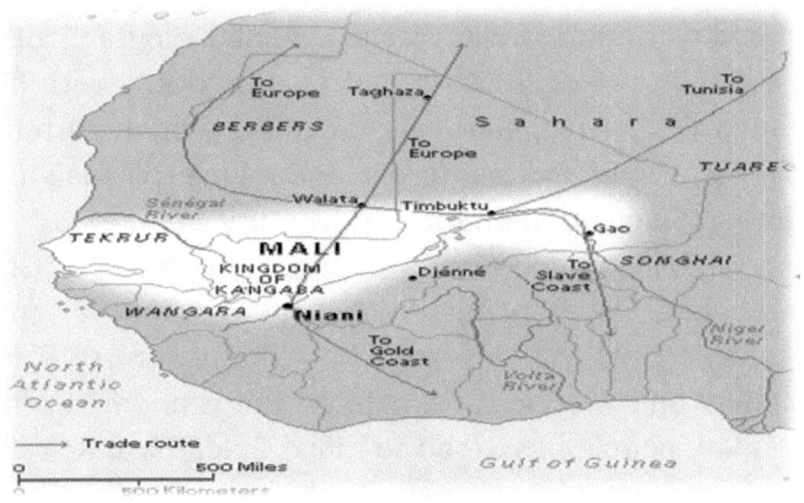

Goods coming from the Mediterranean shores and salt were traded in Timbuktu [Mali] for Gold. The prosperity of the city attracted black scholars, black merchants and Arabs traders from North Africa. Salt, books and Gold were very much in demand at that time. Salt came from Tegaza in the north, Gold, from the immense Gold mines of the Boure and Banbuk and books, were the refined work of the black scholars and scholars of the Sanhaja descent. "The booming economy of Timbuktu attracted the attention of the Emperor of Mali, Mansa Mussa (1307- 1332) also known as 'Kan Kan Mussa.' He captured the city in 1325. As a Muslim, Mansa Mussa was impressed with the Islamic legacy of Timbuktu.

According to Volume IV UNESCO General History of Africa, pages 197-200; "Mansa Mussa's pilgrimage to Mecca had made Mali known worldwide. The great ruler took 60,000 porters with him. Each porter carried 3 kilograms of pure Gold, that is, 180,000 kilograms or at least 180 tons of Gold He had so much Gold with him that when he stopped in Egypt, the Egyptian currency lost its value and as result, the name of Mali and Timbuktu appeared on the 14th century world map.

Musa's generous actions spreading Gold throughout local populations, however, inadvertently devalued Gold for 12 years within the regions of Cairo, Medina and Mecca. It devastated the economy in that ordinary people possessed so much Gold, it lost its value as a medium of exchange. To balance super inflated prices at that time an attempt to adjust the people's newfound wealth was implemented.

Musa's generous actions spreading Gold throughout local populations, however, inadvertently devalued Gold for 12 years within the regions of Cairo, Medina and Mecca. It devastated the economy in that ordinary people possessed so much Gold, it lost its value as a medium of exchange. To balance super inflated prices at that time an attempt to adjust the people's new found wealth was implemented.

> *"To rectify the Gold market, Musa borrowed all the Gold he could carry from money-lenders in Cairo, at high interest. This is the only time recorded in history that one man directly controlled the price of Gold in the Mediterranean."*[16]

Here it is clear that not only was Gold employed by ancient civilizations of Arabia eons ago, but Black men of West Africa once controlled the Gold market to the good. The problem was Egypt's economists were not advanced enough to handle that much wealth during the 13th century to accept a transfer of wealth! Whereas the original Black nation ultimately lost his nation's Gold horde to Caucasoid enterprises of London, during these end times, GOLD has come to benefit none in that we are now living in the fall of AN OLD WORLD OF DOING BUSINESS.

The Caucasoid has and is perpetrating a fraud with the amounts of Gold they claim in their Central bank

---

[16] http://en.wikipedia.org/?title=Mansa_Musa

vaults to justify printing paper money, issue credit and lending at criminal usurious rates.

China Invented Paper Money

China's Northern Song Dynasty had their opportunity to rule between 960 and 1279. This people were the first government in world history to issue banknotes or paper money. Incidentally, Marco Polo upon his journeys throughout China brought back to the West (Europe) the first stories of paper money in China. But little did he know, the idea ultimately led to ruining China's experimental economy.

"During the Song Dynasty, the merchant class became more sophisticated, well-respected and organized than in earlier periods of China. Their accumulated wealth often rivaled that of the scholar-officials who administered the affairs of government. For their organizational skills, Ebrey, Walthall, and Palais state that Song Dynasty merchants...set up partnerships and joint stock companies, with a separation of owners (shareholders) and managers. In the large cities, merchants were organized into guilds according to the type of product sold; they periodically set prices and arranged sales from wholesalers to shop owners. When the government requisitioned goods or assessed taxes, it dealt with the guild heads."[17]

History proves ancient Northern Chinas large government run industries and large privately-owned business enterprises were more sophisticated than the ones preceding it. Nonetheless, these concepts were

---

[17] http://en.wikipedia.org/wiki/Song_Dynasty

expansions of the economic practices of Chinas more ancient Black inhabitants who modeled government before the "Mongoloid" yellow race of China came into power.

In the book BLACKS IN ANCIENT CHINA by Clyde Winters he states "In northern China the blacks founded many civilizations. The three major empires of China were the Xia Dynasty (c.2205-1766 B.C), Shang/Yin Dynasty (c.1700-1050 B.C) and the Zhou Dynasty. The first dynasty of China was Xia (She-ya). The Xia civilization of ancient China lasted from 2205 to 1766 B.C. According to the Guben zhu Shu zhi Nien, the Xia dynasty "from Yu to Zhieh had seventeen kings... and lasted 471 years". Clyde Winters goes on to write: "Here the soil was fertile and black Chinese farmers grew millet 4000 years ago, and later soybeans. They also raised pigs and cattle. By 3500 B.C., the blacks in China were raising silkworms and making silk. The Africans or blacks that founded civilization in China were often called li min "black headed people" by the Zhou dynasts. This term has affinity to the Sumero-Akkadian term sag-gig-ga "black headed people". These li min are associated with the Chinese cultural hero Yao."

When Chinas ancient black civilizations were displaced from their seats of power and authority, the yellow man (Mongoloid race) of China carried forth experimentally, the black man's invention of government and business affairs along the Yellow river.

For instance, when China's Song Dynasty emerged

to dominate the market system, they established thousands of small private businesses and entrepreneurs. Rather in large suburbs and rural areas, which thrived off the economic boom of the period, Song was the place to conduct business. There was also room for small economic success with the "inn keeper, the petty diviner, the drug seller, the cloth trader," and many others.[18]

Do you see any resemblance and comparison between modern western commercial nations and the heights to which China's ancient Song Dynasty had reached in business types? Well, it's no accident. As an old expression says, "there is nothing new under the sun".

Another great sign left by black Africa's presence in China is demonstrated by the images of the Xia dynasty army below. These soldiers were carved in stone as a reminder of their greatness and

[18] http://en.wikipedia.org/wiki/Economy_of_the_Song_Dynasty

organizational genius.

China has about 5,000 years of continuous history, but early 'records' are of a mythological and legendary nature. The Xia Dynasty, documented in early Chinese histories as the first Chinese dynasty, dates back to 2,200 BC.

The next photo below (l) is of a Chinese soldier of the Qin dynasty army with Africoid (Black) features and the photo next to it is of the late Dr. Khalid Muhammad, founder the of the New Black Panther Movement.

The Qin soldier image as well as thousands of other hidden treasures in the Far East demonstrates an early Black nation presence among the yellow race (Mongoloids). Regrettably, as anti-black sentiments began to dominate the global political landscape, the truth about the Black father of civilization was rewritten, repainted and diminished.

In any case, the Chinese seems to invent the first metal coins before 900 BC, in a tomb near Anyang among the Shang [black civilization] Dynasty.

> "At that time, the coin itself was a mock of more earlier used cowry shells, so it was named as Bronze shell…. By 1000 unification was complete and China experienced a rapid period of economic growth…Song merchants rapidly adopted forms of paper currency starting with promissary notes in Sichuan called 'flying money' (feiqian). These proved so useful the state took over production of this form of paper money with the first state-backed printing in 1024. By the twelfth century various forms of paper money had become the dominant forms of currency in China and were known by a variety of names such as jiaozi, qianyin, kuaizi, or guanzi…

> "The **Mongol-founded Yuan dynasty (Chinese, 1271–1368)** also attempted to use paper currency. Unlike the Song dynasty they created a unified, national system that was not backed by silver or gold. [Then came the Yuan Dynasty].

> "The currency issued by the Yuan was the world's first fiat currency, known as Chao. The Yuan government attempted to prohibit all transactions in or possession of silver or gold, which had to be turned over to the government. Inflation in 1260 caused the government to replace the existing paper currency with a new paper currency in 1287, but inflation caused by undisciplined printing remained a problem for the Yuan court until the end of the Dynasty."[19]

---

[19] http://en.wikipedia.org/wiki/Chinese_currency#Ancient_currencies

As you can see counterfeit money—not supported by Gold or Silver—did not work among the yellow Mongoloid Yuan Chinese and it is failing among the Caucasoid white race who acquired the idea from China. I repeat if there is any resemblance and comparison between ancient Chinas monetary policy and the Western world's monetary policy, it is no accident. As the old expression says, "there is nothing new under the sun." The short end of the story is Caucasians of Europe only borrowed what older cultures had rehearsed within the past 6,000 years.

After creating the monetary foundation for this current world monetary system, the white races future was contained, limited and preprogrammed to self-destruct. It's beginning and ending was vouchsafed by the ancients. An old scripture said, *"The day of judgment is here; your destruction awaits! The people's wickedness and pride have reached a climax. Their violence will fall back on them as punishment for their wickedness. None of these proud and wicked people will survive. All their wealth will be swept away. Yes, the time has come; the day is here! There is no reason for buyers to rejoice over the bargains they find or for sellers to grieve over their losses, for all of them will fall under my terrible anger. And if any merchants should survive, they will never return to their business. For what God has said applies to everyone—it will not be changed! Not one person whose life is twisted by sin will recover."(Ezekiel 7:10-13)*

China ultimately lost power under its Qing Dynasty to British demands for trade exchange advantages. Low Chinese demand for westernized goods, and high westernized demand for Chinese goods, including tea, silk, and porcelain, forced westernized merchants to purchase Chinas products with silver, the only commodity the Chinese would accept. During that era, such demand put the British empire at a disadvantage because Britain was employing the gold standard. And since China only would accept silver, Britain had to purchase silver from other European and/or Western countries using its gold reserves. Of course, this only made matters worse for transaction cost to purchase Chinese goods.

To solve this problem, the British begin exporting opium (drugs) to China that ultimately led to the Anglo-Chinese Wars. Eventually open warfare between Britain and China broke out in 1839.

"China was defeated in both wars leaving its government having to tolerate the opium trade. Britain forced the Chinese government into signing the Treaty of Nanjing and the Treaty of Tianjin, also known as the Unequal Treaties, which included provisions for the opening of additional ports to unrestricted foreign trade, for fixed tariffs; for the recognition of both countries as equal in correspondence; and for the cession of Hong Kong to Britain. The British also gained extraterritorial rights. Several countries followed Britain and sought similar agreements with China. Many Chinese found these agreements humiliating and these

sentiments contributed to the Taiping Rebellion (1850–1864), the Boxer Rebellion (1899–1901), and the downfall of the Qing Dynasty in 1912, putting an end to dynastic China.

"Another name for this rebellion is **"The Society of Righteous and Harmonious Fists** *or* **"Boxers United in Righteousness".** *It was also during this uprising when white Christians in China were re-classified as the Devil.*[20]

When the rebellion ended, members of the **'Society of Righteous and Harmonious Fists"** were captured and/or killed. On the next page I have provide an actual photo taken to display how some of them looked and please do not be surprised!

---

[20] http://en.wikipedia.org/wiki/Boxer_Rebellion

Some of China's trouble-makers—"Boxer" Prisoners captured and brought in by 6th U.S. Cavalry—Tientsin, China. Copyright 1901 by Underwood & Underwood.

Image source: www.dataprocess.org/rdswk/boxer-rebellion.html

Chapter FIVE

Merchant Jews Of Norman England

Over population into Scandinavia during year 700 C.E. compelled the "Vikings" of Sweden, Norway, and Denmark to begin making their oversea excursions into Ireland, England, Scotland, France, and elsewhere.

Normans were descended from Danish Vikings of Scandinavia. This Caucasoid tribe knew the weaknesses of their brethren from the days of raiding other mainland tribes whom were colonized in wide areas of Europe. The earliest documented raids by the Vikings began in 793 at Lindisfarne, England. The first phase of attacks were from 790-840. The second phase of Scandinavian activity occurred from 841-875. Vikings arrived, unexpectedly, by plundering, burning, killing or enslaving the inhabitants and then leaving the conquered lands. They met little organized resistance...In the third phase between the years of 876-911, the Vikings, along with their Great Army, continued to plunder on both sides of the English Channel and began to colonize England and France. They also permanently settled in lands they had raided such as Ireland, Iceland, and areas in Russia around Novgorod and Kiev. Charles the Simple, king of the West Franks, ended the Viking raids in 911 by giving Normandy, France to the Vikings.[21]

---

[21] www.thenagain.info/WebChron/WestEurope/VikingRaids.html

In 1066, William the Conqueror of Normandy, Scandinavia finally invaded and conquered England. When William arrived he brought with him Naphtali Jewish merchants who were employed to help Norman Christians to conquer all English feudal occupied lands for the purpose of collecting taxes to the royal treasure. (See image of William below)

The merchants were in the "exclusive domain of the King's personal control" and also were traffickers in Torah traditions. Politically they could only issue loans on behalf of the king and other Royals.

History records Christian kings relegating Jewish merchants to lend money for interest because usury is permitted according to the Jewish Babylonian Talmudic tradition under specific circumstances while Christianity forbids usury. Tradition or not, the Jewish system of usury drove the early gentile nations into madness. While one hand washed the other, leaving none clean, the All Seeing Eye of God was well aware of their financial crimes.

"Unique among its feudal neighbors, the Norman Duchy was governed as a centralized unit...In Normandy...feudalism was strictly territorial: a pyramid of land tenure embodied a system of military obligations ascending from knight through baron to Duke, from whom all land and authority derived. On the continent, and later in England, William the Conqueror set out to maintain and strengthen this Norman system of centralized governance. With the Conquest, the

Normans introduced to England a well-organized central authority. The early governance of conquered England concentrated power in the King. As William the Conqueror imposed the rigorous order of the feudal system, he avoided the system's tendency toward decentralization and disintegration that had sapped the power of the French kings. He limited the power of his tenants-in-chief by granting each of them landholdings scattered over the realm, instead of large, contiguous tracts. He governed the counties through sheriffs who depended on him for their power. He maintained a national militia, thereby shunning total reliance on the loyalty of his tenantsin- chief. And he had all significant landholders swear an oath of primary allegiance to him. This concentration of power in the monarch grew during the successive reigns of a series of strong kings who increasingly assumed more power— military, legislative, and judicial—over the nation...

"The Jews in Norman England, however, were within the exclusive domain of the King's personal control, living at his sufferance and according to his wishes. The first settlement of Jews in England came in 30 the wake of William the Conqueror. William determined that he should be the sole owner of Jews in England. Others could own Jews only with the King's permission as expressed by royal grant...The underlying reality was that the Jews were no more than the embodiment of the King's accounts receivable. Jews were subject to periodic tallage and tithing when the King required them to turn over money that was held, ultimately, on his behalf. The King preserved the Jews and their investments as representing his own financial future.

"The royal charters, in effect, permitted the Jews usufruct of money [The right to use and enjoy the profits and advantages of something belonging to another as long as *the property is not damaged or altered in any way] much as their Christian neighbors were permitted use of the land. At the King's pleasure, they would derive a livelihood by lending money at interest. Because Jews could lend money at interest, they were available to finance excursions to continental Europe and on Crusade. In addition to the extraordinary fiscal demands of the Crusades [against the Islamic world], the nobles still owed knight service.*

"Taxpaying began to replace personal service in the practice of "scutage"—money assessed from landowners in lieu of knight fees. For this too, the Jews' assets were liquid, and available for a fee. It was convenient to the realm to have a source of credit. It was further convenient that the profits from the loan arrangements, forbidden to Christians, be available to the King via his Jews. And it was to the King's advantage to enforce the contracts of credit made by the Jews".[22]

In what way the Jews of Europe practiced their business dealings, discontent built up against them. By 1290 they were expelled from England.

"Persecution of England's Jews could be brutal, recorded are deadly massacres at London and York during the crusades in 1189 and 1190.

"On November 17, 1278, all Jews of England, believed to have numbered around 3,000, were arrested on suspicion of coin clipping and counterfeiting, and all Jewish homes in England were searched. At the time, coin clipping was a widespread practice, which both Jews and Christians were

---

[22] www.freedom-school.com/admiralty/how-jewish-law-became-english-law.pdf

involved in, and a financial crisis resulted, and according to one contemporary source, the practice reduced the currency's to half of its face value.

"In 1275, coin clipping was made a capital offense, and in 1278, raids on suspected coin clippers were carried out. According to the Bury Chronicle, 'All Jews in England of whatever condition, age or sex were unexpectedly seized … and sent for imprisonment to various castles throughout England. While they were thus imprisoned, the innermost recesses of their houses were ransacked.' Some 680 were detained in the Tower of London. More than 300 are believed to have been executed in 1279. Those who could afford to buy a pardon and had a patron at the royal court escaped punishment.

"Edward I increasingly showed anti-Semitism as in 1280 he granted a right to levy a toll on the rivulet bridge at Brentford 'for the passage of goods over it, with a special tax at the rate of 1d. each for Jews and Jewesses on horse, 0.5d. each on foot from which all other travellers were exempt.' This antipathy eventually culminated in his legislating for the expulsion of all Jews from the country in 1290. Most were only allowed to take what they could carry. A small number of Jews favored by the king were permitted to sell their properties first. Almost all evidence of a Jewish presence in England would have been wiped out if it had not been for the efforts of one monk, Gregory of Huntingdon, who purchased all the Jewish texts he could to begin translating them.

"From then until 1655, there is nearly no official record of Jews in England with a few exceptions, for example Jacob

Barnet, who was ultimately arrested and exiled."[23]

One may derive from this historical account that members of the Jewish community made self-righteous Christian elites, nobles and kings not only feel superior, but justified in taking profits (taxes) earned by usury from their citizens as long as Jewish merchants (lenders) handled the transactions. However, as history records, such an unholy alliance was short lived.

Jews were readmitted back into England's public arena in 1655, and, by 1690, about 400 Jews settled. The British Isles would ultimately become a base of operation from where an unholy alliance between Jews and Anglo Saxon Signore (Banking) families would govern world affairs. Their rulership would be based upon their biblical interpretation and action to fulfill destiny up to the time when their world crumbles.

> *"Jew and Gentile are the same in this respect. They all have the same Lord, who generously gives his riches to all who ask for them." (Romans 10:12)*

European (White) Jews

Both the Gentile and Jewish Caucasians origin of reality begins in the Caucus Mountain Range of Europe—West Asia. Those whom had first followed Moses (Musa) were called Jews, according to the teachings of the Honorable Elijah Muhammad, and so they were the first to leave the Caves and made their way into the ancient black towns, villages and cities around Asia Minor.

"Anatolia (from Greek Ἀνατολή, Anatolḗ — "east" or "(sun)rise"), in geography it is known as Asia Minor (from Greek: Μικρὰ Ἀσία Mīkrá Asía ("small Asia"); in modern Turkish: Anadolu), Asian Turkey, Anatolian peninsula, or Anatolian plateau, denotes the westernmost protrusion of Asia, which makes up the majority of the Republic of Turkey.The region is bounded by the Black Sea to the north, the Mediterranean Sea to the south, and the Aegean Sea to the west."[24]

Before migrating out the Caves over 4000 years ago, one key aspect of what Prophet Moses (Musa)[25] taught to their ancestors was clean up, how to use money, keep a low profile (blend among the nationals) and do not join the religion (way of life) of the original civilization. Moses also taught the former cave dwellers how to defend self with weapons (Deuteronomy 33:20).

The Holy Quran chapter 18—The Cave—provides cryptic verses which reads: *"18:18 And thou mightest think them awake while they were asleep, and We turned them about to the right and to the left, with their dog outstretching its paws at the entrance (**of the cave**). If thou didst look at them, thou wouldst turn back from them in flight, and thou wouldst be filled with awe because of them. 18:19 And thus did We rouse them that*

---

[24] http://en.wikipedia.org/wiki/Asia_Minor
[25] Zul-qarnain and Moses' story are both mentioned in this chapter for a reason. You can draw your own conclusions.

they might question each other. A speaker from among them said: How long have you tarried? **(in the caves)** They said: We have tarried for a day or a part of a day. (Others) said: Your Lord knows best how long you have tarried. **Now send one of you with this silver (coin) of yours to the city**, then let him see what food is purest, and bring you provision from it, and let him behave with gentleness, and not **make your case known to anyone**. 18:20 For if they prevail against you, they would stone you to death **or force you back to their religion**, and then you would never succeed." (Holy Quran 18:18-20) On the next page is an ancient map of Asia Minor. The black dots represent Jewish settlements. Of course you may not recognize the ancient names scene on the map.

Researchers have discovered city remains of ancient black civilizations of Asia Minor as old as 9,000 year old.

> "The main cities of Hatti were, Mahmatlar, Horoztepe, Alacahoyuk and Hattus. Their religion was nature/animalistic as their gods depict various aspects of nature in the form of animals."[26]

Caucasian Jews did not enter into Asia Minor— Jerusalem in Palestine, at all, until around 4000 years ago. The black man called this city Jebus; also Salem and Ariel.[27]

If one were to travel throughout modern day

---

[26] http://www.realhistoryww.com/world_history/ancient/Anatolia_Turkey.htm
[27] Supreme Wisdom Lesson No. 1 of the Nation of Islam in the West

Turkey today, none of the ancient aboriginal cities, towns and villages existed. Nevertheless, it was black people whom gave language to the white race or [indo Europeans] starting also around 4000 years ago.

"Prof Mark Pagel, a Fellow of the Royal Society from the University of Reading who was involved in earlier published phylogenetic studies, said:

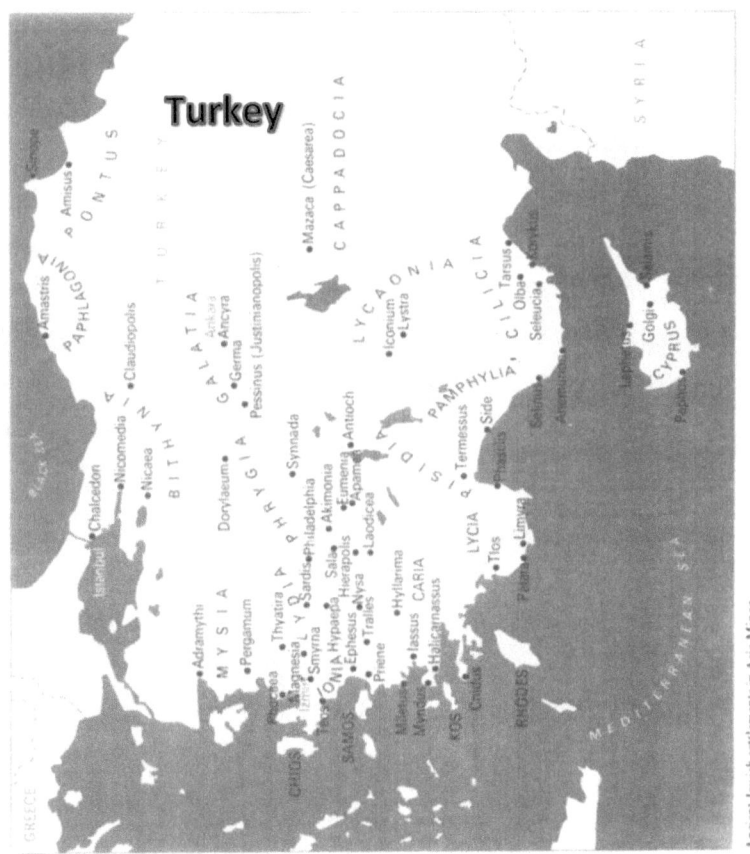

Ancient Jewish settlements in Asia Minor.

'This is a superb application of methods taken from evolutionary biology to understand a problem in cultural evolution - the origin and expansion of the Indo-European languages.

"This paper conclusively shows that the Indo-European languages...arose, as has long been speculated, in the Anatolian region of what is modern-day Turkey and spread outwards from there." This also includes the English language which has its roots in what developed in ancient Asia Minor—a small civilization or off-shoot of the more complex ones originated by the black nation of the earth!

From out of the caves and after a few thousand years fighting against the original black nation and then fighting among themselves i.e. Vikings against eastern and western Europeans and Greeks verse the Romans and Romans against the Jews and Germanic tribes, the Caucasian race were able to produce their first Gold coins.

Chapter SIX

Europe's Golden Coins Of 1252

Peter Spufford, *Money and its Use in Medieval Europe* (Cambridge, 1986) writes: "The earliest were the Florentine fiorino d'oro or florin, worth 240 Florentine denari or 20 Florentine grossi and the Genoese genovino, worth 96 Genoese denari. Both began to be issued in 1252. These were coinages made of Gold imported from Africa. In the second half of the thirteenth century the use of Gold coins in Europe spread only slowly outside northern Italy. The opening up of Gold mines in Europe itself in the first half of the fourteenth century, principally at Kremnica in the kingdom of Hungary, caused the use of Gold coins in Europe to spread much more rapidly. By the middle of the fourteenth century Gold coinages were established in France, the Low Countries, England and the Rhineland, as well as Hungary, and the Gold currency of Italy and Spain had become much more abundant".

During the mid-thirteenth century regular Gold coins were issued outside Muslim-influenced Spain and Sicily, Italy. **Europe was also evolving out of its heathen dark age thanks to Islam**. Therefore out of northern Italy emerged the Lombard bankers during the 13th and 14th century. These people derived from a Germanic Teutonic tribe of Vikings that moved westward from East Scythia into the Scandinavian Peninsula from southern Sweden and gradually worked their way into

Mediterranean apt Italy. While in Italy, eventually everyone spoke Italian became Catholic and Italy became another Germanic area.

> *"Perhaps most importantly, the Lombards got involved in political arguments with the Pope, and this was what caused the papacy to call upon the Franks for aid. The papacy was a prize every medieval magnate wanted to possess. But the popes knew that they could not long survive if they were the creature of one king or emperor.*

> *"The Moslems had conveniently removed the authority of the Eastern Roman emperor from Italy (with a little help from the Lombards), but someone was needed to keep the Germans in Italy (and elsewhere) from controlling the papacy. For several centuries the protector of the papacy became the Franks (and later the French)."[28]*

Historians admit Lombard creative accountancy enabled them to avoid the Christian sin of usury wherein interest form a loan was [re] recorded in the accounts either as a voluntary gift from the borrower or as a reward for the risk taken. Is this system also where the U.S. Internal Revenue borrowed its voluntary compliance law whose definition is:

> *"A system of compliance that relies on individual citizens to report their income freely and voluntarily, calculate their tax liability correctly, and file a tax return on time," according to the Internal Revenue Service.*

> *"The income tax system is voluntary. That's because people*

---

[28] www.hyw.com/Books/History/Langobar.htm

are free to arrange their financial affairs in such a way to take advantage of any tax benefits. Voluntary does not mean that the tax laws don't apply to you. Voluntary means you can minimize your taxes by taking advantage of various deductions and tax credits."

"Voluntary also means that you must tell the IRS what your tax liability is. And the only way to do that is to file a tax return."[29]

Thus, we see right away, the fraud of usury within Western Europe's monetary scheme against its populous as early as the 13th century. Did this "creative" double-entry bookkeeping re-recorded as a voluntary gift derive from ancient Islamic and righteous scholars and scientists of the Black nation? Of course not! The wise Allah people and scientists of Islam and religion in general attempted to reveal the dangers of unregulated usury. They had it written in the Holy Quran—Chapter 2:275? Or did Lombard bankers single handedly devour usury among the heathens of Western Europe beyond its original intent and limitations?[30]

"Those who devour usury will not stand except as stand one whom the Evil one by his touch Hath driven to madness. That is because they say: "Trade is like usury," but Allah hath permitted trade and forbidden usury. Those who after receiving direction from their Lord, desist, shall be pardoned for the past; their case is for Allah (to judge); but those who

---

[29] http://taxes.about.com/od/taxglossary/g/Voluntary.htm
[30] www.hyw.com/Books/History/Langobar.htm   Study or collection of money, coins, medals, tokens, exonumia and banknotes.

repeat (The offence) are companions of the Fire: They will abide therein (forever)."(Holy Quran 2:275)

I reiterate, the United States Internal Revenue Service (IRS) seems to have taken a page out of the Lombard's playbook of creative accountancy as it deals with collecting U.S. income taxes. The IRS penalties, as set out in Section 1441 Internal Revenue Code, are imposed to "**enhance voluntary compliance.**"

Without delving into a great conspiracy theory surrounding U.S. IRS tax collection jurisdiction, let us just quote a verse from the Holy Quran wherein Allah and the Allah people say: *"As for those who disbelieve: 'Were not My revelations recited to you, but you turned arrogant and were a criminal people?'" (Quran 45:31)*

Early founders of Caucasoid banking procedures were more about collecting interest, by taking advantage of another's misfortune and/or ignorance, through lending We know from earlier quotes in this book, Lombard's were of the Germanic strain of the Caucasian race that had invaded and established themselves in Italy—southern Europe.

"Lombard banking refers to the historical use of the term 'Lombard' for a pawn shop in the Middle Ages, a type of banking that originated with the prosperous northern Italian region of Lombardy (hence the name).The term was sometimes used in a derogatory sense as townspeople often accused lombards of usury. The Lombards were never very popular for this reason and often had trouble with local governments. They were careful to write down all of the

terms of contracts, to protect themselves, but that was no guarantee for success in times when few others could read and write, including local leaders."[31]

After the 13th century Lombard bankers of Florentine set the stage for pawnshop banking, there evolved in Italy a few more powerful bank families during the 14th century; namely the Bardi and the Peruzzi. These two families grow their wealth by offering financial services mainly to the papacy (Popes of Rome). They also earned great sums of money charging bank fees for exchanging bills for business merchant traders. The Bardi family established bank offices outside of Italy; namely, in Barcelona, Seville and Majorca, in Paris, Avignon, Nice and Marseilles, in London, Bruges, Constantinople, Rhodes, Cyprus and Jerusalem. Theses Lombard's were so powerful that historians blame them for Europe's first major "Global" financial crash 650 years ago. The reason given: Europe's illiterate rulers were heavily in debt due to war loans owed to Florentine privatized bank networks of the Lombard's and were unable and unwilling to pay interest fees.

Take for instance Edward III of England, in year 1340. He borrowed 600,000 Gold florins from the Peruzzi and another 900,000 to wage war against France. Some historians contest Lombard bankers even forced loans upon borrowers. So in 1345 Edward III of England defaulted on his payments, reducing both

---

[31] http://en.wikipedia.org/wiki/Lombard_banking

Florentine houses of the Lombard's to bankruptcy. Nevertheless, Florence as a great banking centre even survives the disaster because they created a counter plan. (1) remove the Crowns (government of the king) strict monopoly on the coinage and money supply and (2) circumvent Western Christianity's cannon law which outlawed usury, particularly on non-governmental merchant capitalism or private commerce insurance for financiers.

Thereafter, the next great bank family to evolve out of the financial meltdown of 1345 were the prominent 15th century Pazzi and the Medici.

*"The Medici family was a powerful and influential Florentine family from the 13th to 17th century. The family produced three popes (Leo X, Clement VII, and Leo XI), numerous rulers of Florence (notably Lorenzo the Magnificent, patron of some of the most famous works of renaissance art), and later members of the French and English royalty. Like other Signore families (**government; governing authority families with sovereign powers or lordship over the land and people and sea**) they dominated their city's government. They were able to bring Florence under their family's power allowing for an environment where art and humanism could flourish...The Medici Bank was one of the most prosperous and most respected in Europe. There are some estimates that the Medici family was, for a period of time, the*

wealthiest family in Europe. From this base, the family acquired political power initially in Florence, and later in wider Italy and Europe."[32]

Again, by necessity without the blessing of the Christian fathers of the medieval Western Roman Catholic Church, Europe's privatized insane banking networks and merchants understood they had to get around (1) the usury doctrine to 'disguise' interest payments in the exchange insurance commercial rate and (2) the almost universal bans on bullion (Gold) exports, especially when trading internationally with the Islamic world. The private merchant class knew these restrictions represented boundaries in which they could not accept since they needed to saddle all European Crowns and the globe for that matter to perpetuate their financial crimes.

Therefore, maddened private financiers figured out how usury could be applied into a form of insurance since they took chances on investments, which were once outlawed by the Holy See—Christian world.

"Insurance is chancing ['relying on or inviting the risks of chance' – Merriam-Webster's Unlimited Dictionary] in the sense that the premiums are paid for certain, whereas the return is uncertain. You may lose all the premiums you paid or may receive in return more then what you paid. This is known as chancing. There is also interest as money is being exchanged here for money and one party pays less and receives more in return. Hence: "This is the reason why the

---

[32] www.lumrix.net/health/Medici.html

great contemporary [Islamic] scholars from all over the world have declared all types of prevalent insurances unlawful (haram), unless when one is compelled to effect it by the Government."[33]

Here again, the Allah people of the Islamic world, understood long ago how and why misapplied insurance and usury is prohibited because it induces financial injustice that is a violation of the sacredness of the people. *Pretend value* compared to *true value* destroys actual purchase power. After all, what else is more sacred to a man or women other than our purchasing power and religious faith?

In cases of Islamic government-compelled ventures, loans are secured by a nation's taxes. So why and when did the Christian world surrender its foundation. According to John H. Munro wrote in his abstract, The Late-Medieval Origins of the Modern Financial Revolution: Overcoming Impediments from Church and State he writes: *"...from the late thirteenth century, the Crown was incorporating the then evolving international Law Merchant [Bankers rules] into statutory law, and it also established law merchant courts...In 1436, a London law-merchant court was the first, in Europe, to establish the principle that the bearer of a bill of exchange, on its maturity, had full rights to sue the 'acceptor' or payer, on whom it was drawn, for full payment and to receive compensation for damages.*

---

[33] http://qa.sunnipath.com/issue_view.asp?HD=1&ID=383&CATE=43

From that precedent, and then from those provided by similar law-merchant court verdicts in Antwerp and Bruges (1507, 1527), the Estates General of the Habsburg Low Countries (1537- 1541) produced Europe's first national legislation to ensure the full legal requirements of true negotiability –including the right to sue intervening assignees to whom bills had been transferred in payment. These Estates-General also legalized interest payments (up to 12%..."

By the 1600's the Central Bank of England was founded by Europe's

> *"The Bank of England (formally the Governor and Company of the Bank of England) is, despite its name, the central bank of the whole of the United Kingdom and is the model on which most modern, large central banks have been based. It was established in 1694 to act as the English Government's banker, and to this day it still acts as the banker for Her Majesty's Government. The Bank was privately owned and operated from its foundation in 1694. The Bank's headquarters has been located in London's main financial district, the City of London, on Threadneedle Street, since 1734."[34]*

This victory for the Banking and/or merchant elites of Europe legislatively grafted their system into a quasi-government. In so doing, they instituted their financial jurisdiction as the new-land-lord and monetary policy rulers. The more loans they issued meant more money-printing and usury-interest collections. The more

---

[34] http://en.wikipedia.org/wiki/Her_Majesty%27s_Government

money printing also meant more inflation passed down to the populous--consumer. In practice, the term monetary inflation is used to specifically refer to an increase in the money supply.

When the price levels rise, each unit of currency buys fewer goods and services; consequently, inflation is also an erosion of the purchasing power of money – a loss of real value in the internal medium of exchange and unit of account in the economy.

How else was Western Europe going to be saddled and made into a civilization to fulfill their role in scripture? *"[18:94] They said, "O Zul-Qarnain, Gog and Magog are corruptors of the earth. Can we pay you to create a barrier between us and them?" [18:95] He said, "My Lord has given me great bounties. If you cooperate with me, **I will build a dam between you and them.** [18:96] "Bring to me masses of iron." Once he filled the gap between the two palisades, he said, "Blow." Once it was red hot, he said, "Help me pour tar on top of it." [18:97] Thus, they could not climb it, nor could they bore holes in it. [18:98] He said, "This is mercy from my Lord. **When the prophecy of my Lord comes to pass, He will cause the dam to crumble. The prophecy of my Lord is truth."** [18:99] At that time, we will let them invade with one another, then the horn will be blown, and we will summon them all together. [18:100] We will present Hell, on that day, to the disbelievers. [18:101] They are the ones whose eyes were too veiled to see My message. Nor could they hear. (Holy Quran 18:94-101)*

One of the main reasons the Caucasian world did not ever want non-whites reading the Holy Quran is because the book reveals their beginning, ending and origin in divine historical language more clearly than the Bible. The divine history of the Jews and Gentiles of Europe is totally exposed in the Holy Quran. However, in the research language of their scholars, they are exposed as well.

For instance, who are the Nephtali? Who are the Huns and Goths and Swedes, and Visigoths and Cimmerians and Scythians? Did Jews of the Caucasus Mountains marry into these ancient nomadic tribes of Europe?

"Norway was settled by groups (such as the Nephtalite Huns) who mainly descended from Naphtali. Elements from Benjamin, Gad, and other Tribes were also important. The Goths (from Gad) ruled over Norway for some time, as did the Swedes (who are also descended from Gad) after them.

"The symbol of Naphtali was a stag or deer and a deer was a symbol of Scandinavian (including Norwegian) royalty. The Norwegian coat of arms depicts a lion bearing an ax. A lion was one of the symbols of Gad (Deuteronomy 33; 20).

"Moses predicted that a blow from one of the weapons of Gad would be sufficient to sever arm with the head (Deuteronomy 33; 20) indicating the favored use of a striking ax-like weapon as compared with a thrusting pointed one. The ax or something like it was one of the major weapons of the ancient Cimmerians and of the Massagetae who were Goths east of the Caspian and with

*whom the **Nephtalite Huns** were affiliated."[35]*

Notice the name Nephtalite? Nephtali is explained in the following quote by author, Davidy who writes..."*'Some believe them to have become part of the Khazar federation.'* At all events the bulk of the Naphtali had begun previously to migrate westward before 450 c.e. and had disappeared from the Scythian area it is hereby proposed that the earlier NAPHTALITES and bulk of the Naphtalite nation who went westward and were since unheard of became the Vikings of Scandinavia especially Norway' (The Tribes, pp.199, 200).

Today the most powerful Caucasian Jewish Signore family i.e., Rothschild chose to use as their original shield of arms represents the above history.

Rothschild Shield of Arms

Normandy Shield of Arms

---

[35] http://en.wikipedia.org/wiki/Inflation

Western Signore families (Shadow government) know exactly who they are and their prophetic role in geo-politics. The blood of Vikings runs through their veins and they are not Semitic at all. Unfortunately, the former U.S. slaves of America cannot and/or are too afraid to decode the language of symbolism to understand who the enemy of the Original Black Nation and the Black God historically represents

> *"The symbol of Naphtali was 'A hind [doe, deer, stag and/or unicorn] let loose' (Genesis 49:21) and a **deer or stag** was used as **a royal symbol by the Kings** of Scandinavia. The stag also seems to have been **a favourite motif amongst Phoenecian and Israelite craftsmen**.*

> *"The Israelite Tribe of Naphtali therefore became the Nephtali-Huns (or Hephtalitesas they are also called) who together with the Dani were once in east Scythia. (see map 1 and Map 2) From east Scythia the Naphtali migrated to Norway and the Danes to Denmark.*

> *"Norway was settled by groups (such as the Nephtalite Huns) who mainly descended from Naphtali. Elements from Benjamin, Gad, and other Tribes were also important. The Goths (from Gad) ruled over Norway for some time, as did the Swedes (who are also descended from Gad) after them. **The symbol of Naphtali was a stag or deer and was a symbol of Scandinavian (including Norwegian) royalty.***

> *"**The Norwegian coat of arms depicts a lion bearing an ax.** A lion was one of the symbols of Gad (Deuteronomy 33; **Moses predicted that a blow from one of the weapons of Gad would be sufficient to sever arm with the head (Deuteronomy 33:20)** indicating the favored use of astriking*

ax-like weapon as compared with a thrusting pointed one. The ax or something like it was one of the major weapons of the ancient Cimmerians and of the Massagetae who were Goths east of the Caspian and with whom the Nephtalite Huns were affiliated[36]"

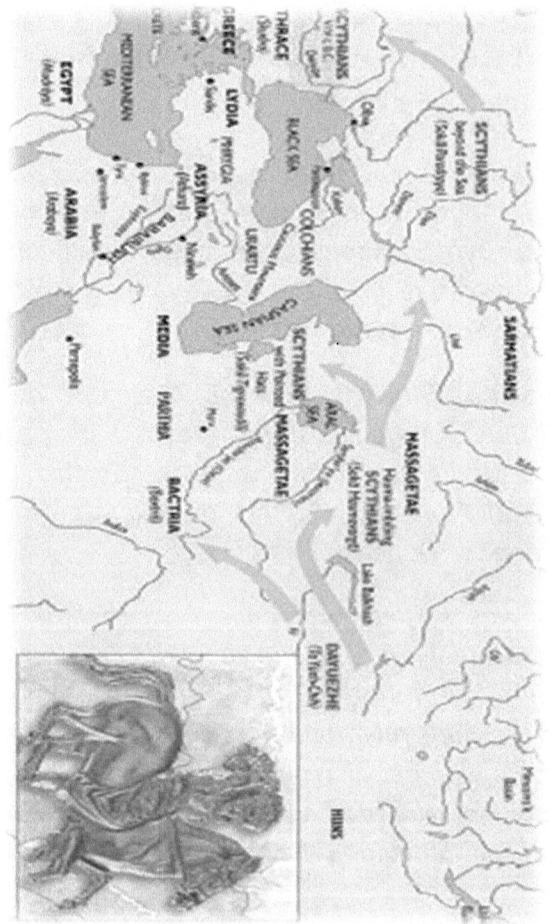

Map 1

---

[36]
www.britam.org/naphtali.html

Map 2

Source: www.raremaps.com/gallery/detail/25913/Tartaria_olim_Scythia/Munster.html

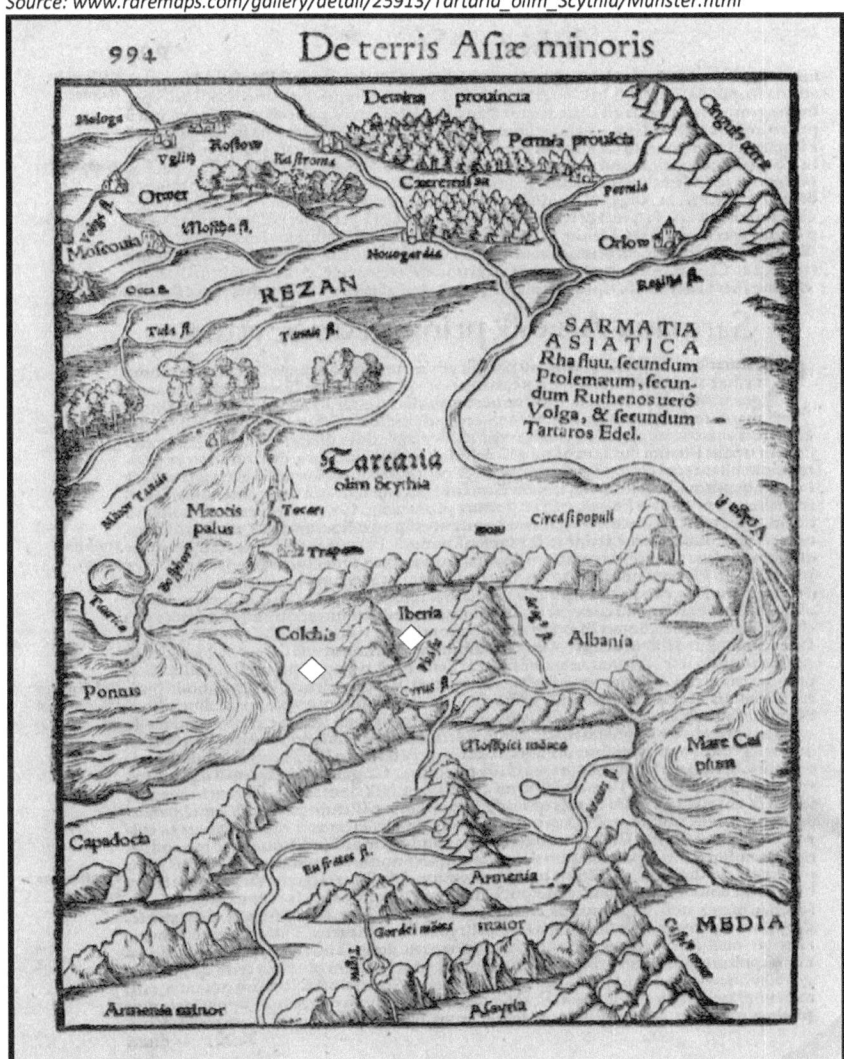

The initial location of where the white race where roped and bound 2000 years in the Caucus Mountains area is called **Iberia**. Colchis represents and old Egyptian outpost.

In this order, use map 2 as a visual guide to see location where the Caucasians dwelled in the Caves of Iberia Europe 2000 years, 2) recall Holy Quran 18:94-101 to understand how the first Caucasian "Jews" migrated out of the caves, and 3) read the Biblical history about the tribe of Dan—the last of Jacobs offspring to be given an opportunity to rule before losing it all.

> "Squeezed into the narrow strip between the mountains and the sea, its energies were great beyond its numbers." Being pressed by the Amorites and the Philistines, whom they were unable to conquer, they longed for a wider space. They accordingly sent out five spies from two of their towns, who went north to the sources of the Jordan, and brought back a favourable report regarding that region. "Arise," they said, "be not slothful to go, and to possess the land," for it is "a place where there is no want of anything that is in the earth" **(Judg. 18:10)**. On receiving this report, 600 Danites girded on their weapons of war, and taking with them their wives and their children, marched to the foot of Hermon, and fought against Leshem, and took it from the Sidonians, and dwelt therein, and changed the name of the conquered town to Dan **(Josh. 19:47)**. This new city of Dan became to them a new home, and was wont to be spoken of as the northern limit of Israel, the length of which came to be denoted by the expression "from Dan to Beersheba", i.e., about 144 miles. "But like Lot under a similar temptation, they seem to have succumbed to the evil influences around them, and to have sunk down into a condition of semi-heathenism from which they never emerged."[37]

---

[37] http://www.bible-history.com/links.php?cat=40&sub=545

Chapter SEVEN

Rothschild Dynasty Fulfill Prophecy

Down the wheel of time, Caucasians of Europe finally produced a great banking Dynasty via Mayer Rothschild and his five-sons—the masters of money. This Jewish family was prewritten in the Bible Luke 16:14-31. In part, their only hope to pull western capitalism from the brink of disaster, during the end times, is to understand the words of the Bible Luke. In this book, the Rothschild's were hidden under the symbolic name "the rich man." He and his brothers were given direction as to what they needed to do on behalf of Lazarus (Dead Black nation of America). More on this later.

Who are the Rothschild's ancestral heroes? Does this family carry any Khazar bloodline or tradition? And why did the Khazar king adopt Judaism in 830 AD rather than Islam or Christianity? Another question to ponder is why did they relocate from Khazaria into the lands of Europe? See map on next page. Was relocating due to what Genesis 10:5 and 10:12 foretold where Caucasian Jews and Gentiles were to establish themselves to seek riches? Gen. 10:5 reads, *"By these were divided the islands of the Gentiles in their lands, every one according to his tongue and their families in their nations."* The islands of the gentiles or Caucasian race are the British Isles.

The island Britain is 800 miles long and 200 miles

broad. There are in the island five nations: (1) English, (2) Welsh (or Britain), (3) Scottish, (4) Pictish, and (5) Latin. The first inhabitants were the Britons, who came from Armenia (Central Asia) and first peopled Britain southward. Then happened it, that the Picts came south from Scythia, with long ships, not many; and, landing first in the northern part of Ireland, they told the Scots that they must dwell there.

Map of Khazar Year 850 AD

At any rate, it was Germany's ruler William IX of Hesse-Kappel who privately consulted with his friend Mayer Amschel Rothschild, a Jewish banker and merchant of Frankfurt. In 1801 he formally appoints him his court agent. By 1803 Rothschild is in a position to lend 20 million francs to the Danish government. Rothschild's Danish loan was one of many transactions he made on behalf of the German government which rapidly establish the Rothschild family as Europe's most powerful bankers.

In fact the Rothschild's ultimately rose to a pre-eminence comparable to that of the Medici and the Fugger in earlier centuries.[38] Mayer Amschel Rothschild and his 5 sons were able to establish bank branches in Vienna, London, Naples and Paris with Great Brittan being the center of their financial universe.

During those days, Napoleon was a French hero and became a Rothschild challenger. He established both the Bank of France and the French bourse (stock exchange), and the National and Departmental Tax Boards to insure equitable taxation for all. By doing so, income of the French peasants skyrocketed. Napoleon's main enemy was the British. He wanted to shape the new world with his ideas and the British had another set of ideas.

"In November 1806, having recently conquered or allied with every major power on the European

---

[38] www.historyworld.net/wrldhis/PlainTextHistories.asp?historyid=ac19

continent, Napoleon issued the Berlin Decree forbidding his allies and conquests from trading with the British...Napoleon was master of continental Europe effectively locking the United Kingdom out of the continent.. It was a large-scale embargo against British trade..."[39]

When Napoleon imposed what was called the Continental System—a blockade aimed at denying the British any trading access to ports in Europe, the bankers, under Rothschild leadership, from behind the scenes, worked their lending skills on all sides. The Rothschild's issued loans of Gold to governments who set forth to defeat Napoleon and his army. Centuries later, in 1969 Time Magazine published: "Rothschild gold bought supplies for the Duke of Wellington before Waterloo." The Battle of Waterloo was fought on Sunday 18 June 1815 in present-day Belgium. It was Napoleon Bonaparte's last battle. He was defeated by combined armies of the United Kingdom, Russia, Austria and Prussia, and an Anglo-Allied army under the command of the Duke of Wellington. By the end of the war the Rothschild family had a vast reputation among their European allies.

---

[39] www.historyworld.net/wrldhis/PlainTextHistories.asp?historyid=ac19

Chapter EIGHT

IMF

The International Monetary Fund was originally created July 1944 with 45 members whose goal it was to stabilize exchange rates and to assist in the reconstruction of the world's international payment system. The payment system is structured to assure all IMF members repay loans they'd receive from its network of central banks including the World Bank.

Another way to put it, the IMF is also a global debt-collection and surveillance agency for money. Countries under its administrative apparatus contribute to a pool which is borrowed by each member, on a temporary basis, by countries with payment imbalances. These transactions include payments for the country's exports and imports of goods, services, and financial capital, as well as financial transfers. It includes an accounting record of all monetary transactions between a country and the rest of the world.

The IMF was especially important when it was first organized to help the Caucasoid civilization stabilize its economic system after many internecine wars. The most solid asset owned by the IMF is Gold at which price it sells cheap to it members. Today IMF membership is 186 countries. It holds 3,217 metric tons (103.4 million ounces) of Gold, which makes the IMF the world's third largest official Gold holder. But the question is: does the

IMF actually "own" the Gold it holds since the bullion belongs, according to IMF articles of agreement adapted at Bretton Woods in 1944, to its member nations.[40]

Bretton Woods

The Bretton Woods system was also established in 1944. Every major capitalist power agreed to "Bretton Woods" economic concepts to initiate a program of national regulation that would prevent the development of nation-state jurisdiction over Central bank jurisdiction in terms of what administrative apparatus would dictate global monetary policy. So a total 730 delegates from 44 countries representing governments and the corporate aristocracy (the Global Elite) deliberated upon and signed the agreements that established a New International Economic Order policed by the International Bank for Reconstruction and Development (IBRD or World Bank), the International Monetary Fund (IMF) and General Agreement on Tariffs and Trade (GATT).[41]

During the Bretton Woods era it was decided to make the American dollar function as the world's dominant currency subsequently leading other nations to believe it would be to their advantage to employ the US dollar first and foremost over what Stalinism in eastern Europe and old Russia were offering i.e. strong arm socialism wherein the government regulates the

---

[40] en.wikipedia.org/wiki/IMF#Organization_and_purpose
[41] www.overlordsofchaos.com/html/Gold___money__bretton_woods_ag.html

capital of the wealthy. So from 1944 to 1971 the U.S. lived by the Bretton Woods provision that US dollars could be redeemed in Gold at the rate of $35 per ounce.

However, as this quasi-one-world-order currency was tested over the years, it failed. None of the other countries Central banks wanted to employ its Gold to back the U.S. dollars global adventures. That is to say, expansion of the global economy was tantamount to more U.S. Gold leaving Fort Knox into the vaults of foreign Central banks abroad because U.S. dollars could be redeemed in Gold at the rate of $35 per ounce. As a consequence, Americas fool hearty increase in US investment abroad, money circulation and military spending meant inflation…inflation…inflation!

Of course, this brought an end to the Bretton Woods Agreement in 1971 and the real price of Gold has been steadfastly suppressed since then. Moreover, Central banks of issue were no longer obliged to redeem their currencies in Gold bullion because their scheme was a success and America became a debtor nation forever.

"Due to limited growth in the supply of gold reserves, during a time of great inflation of the dollar supply, the United States eventually abandoned the gold exchange standard and thus bullion convertibility in 1974. Under the contemporary international currency regimes, all currencies' inherent value derives from fiat, thus there is no longer any thing (gold or other tangible store of value) for which paper notes can be redeemed. One currency can be converted into

another in open markets and through dealers.[42]

In conclusion, Bretton Woods was designed to fail and was only worthwhile to operate while the mass of US dollars circulating in the rest of the world was backed by Gold held in Fort Knox. Although the U.S. holds Gold in vaults, it has not reconstituted its legal right to control its worth in Gold!

Eastern and Western Caucasoid

How are eastern and western elites different? They are different in their objectives to complete global ownership of all earthly resources as possible. One vies for the power of nation-state rule while the other vies for the power of Central bank rule! The Bible refers to these two groups as Gog and Magog. Whereas one side is allied with the Islamic world, the other side is not. It's sort of like "my enemy's enemy is my friend depending upon our business interest."

To bring further clarity to Gog and Magog, I have provided an article written by the Honorable Elijah Muhammad from a February 18, 1972 Muhammad Speaks Newspaper.

> *"Daniel's prophecy of these times, in which we are now living, is equal to the prophecy of John's Revelation (Bible). These two Books of the Bible (Daniel and Revelation) are very similar in their prophecy of the same subject, which deals with the great trouble that shall befall the nations of*

---

[42] en.wikipedia.org/wiki/Convertibility

the earth, at the end of the falling of One nation (America). "The conflict between rulers – America (Gog), Russia (Magog), China and the world-regardless to the power of that one nation (the nation can be small or great) it is all pointing toward the one and the same people: America.

"-Let us take a pause right here to say this; When have we seen the kings of Asia and the kings of the north (China and Russia) coming to America to plead with America for agreement to settle, or to appease, their differences?

"Daniel Ch.11 (Bible) prophecies now of setting the rulers of the earth against America. "And in the end of years, they shall join themselves together; for the king's daughter of the south shall come to the king of the north to make an agreement:"(Dan. 11: 6). But the other nations have made their agreement against America. The kings, the rulers have united and made their agreement against the king of the south's daughter (America's coming.) America is referred to as the daughter of the king of the south.

"This means that America is something that has been produced out of a father-like ruler, which can be none other than England. England once ruled the whole world with a strong hand. Now England's daughter, America, takes over, but America is unable to make friends with the rulers of the earth because of the time and the people- the time given to America is too short for her to make friendship and peace in any confidential way, with the people with whom America is in Conflict. It is too late for America, for the end of the years has come and no more time can be extended, for the sake of an enemy to regain his strength, to fight against the innocent people whom he has crushed. Prophecy has a way of using certain material, which is foreign to actual material that he is aiming to bring to the attention of the people.

"This is to keep the one whom the prophecy is against from knowing until the time of the prophecy's manifestation. So it is with the prophecy of Daniel, Ezekiel, Jeremiah and many other's prophecies. It is tobe made manifest, at the end of the prophecy.

"According to the prophecy of Daniel, Chap. 11, verse 6, America shall be given up and those who be with her to try to strengthen her, shall be given up. The helpers of America shall be the losers."

Elijah Muhammad
Messenger of Allah to you all

Chapter NINE

Fall Of Caucasian Civilization

What the white race or Caucasian man has done thus far is fix his civilization on a sure path to self-destruction. During these modern times, such truth is being borne out. Their craving for gold and every other earthly resource, including black human bodies is and has led to the fall of Caucasian civilization.

But soon enough, all humanity shall be free of slave labor and debt slavery as a result of financial violence via Gold manipulation to justify inflation and false profits for the 1 percent and/or 10 percent Signore families or blood suckers of the poor. For case study: On 21 January 1980 the gold fixing reached the price of $850, a figure not overtaken until 3 January 2008 when a new record of $865.35 per troy ounce was set in the a.m. fixing. However, when indexed for inflation, the 1980 high corresponds to a price of $2398.21 in 2007 dollars, thus the 1980 record still holds in real terms.[43] How will the nations of the earth react if they were to discover Gold is really $25,221.93 per/oz based upon its hypothetical (constant) growth rates indexed with inflation and the eroding value of man-made paper money?[44]

As it stands today, all monetary estimates are

---

[43] en.wikipedia.org/wiki/London_Gold_Fixing
[44] www.bibleprophesy.org/goldrates.htm

calculated at the rate of 1oz gold = $1000 US dollars. But if 1oz gold = $25,000 US, no IMF member country would be able borrow. Nor could Central banks lend or lease the yellow metal to bullion banks at 1% or 2% to invest in government bonds for profits.

As a matter of fact, were it not for The London A.M. Gold Fixing Gold prices would be sky high. Therefore, the real price of Gold is suppressed by the London Fixing to keep their civilization in control of commercial nations and the people thereof for two reasons:

"1) Suppressing the price of Gold has made it a cheap source of capital for New York bullion banks, which borrow it for as little as 1 percent of its value per year. This means Gold is borrowed from central banks and sold, and the proceeds are invested in the financial markets in securities that have much greater rates of return. As long as the price of Gold remains low, this "Gold carry trade" is a financial bonanza to a privileged few at the expense of the many, including the Gold-producing countries, most of which are poor. If the price of Gold were allowed to rise, the effective interest rate on Gold loans would become prohibitive.

"2) Suppressing the price of Gold gives a false impression of the U.S. dollar strength as an international reserve asset and a false reading of inflation in the United States. Too much Gold is being consumed at too cheap a price. Massive amounts of derivatives are being used to suppress the Gold price. If this situation is not corrected soon, there will be a Gold derivative credit and default crisis of epic proportions that will threaten the solvency of the largest international banks and the world standing of the dollar. The Gold Anti-Trust Action Committee requests that a full and complete investigation be launched into this matter as soon as possible.[45]

As state earlier in this book, International Gold trading is an around-the-clock business, with the London Market overlapping those of the Far East in the morning and New York in the afternoon. London bullion traders can make deals from about 7:15 a.m. So what the white race or Caucasian Signor Families are perpetrating is the greatest Ponzi scheme every known to man. The so-called white race abuse of the intrinsic value of Gold has disqualified them to rule the kingdom (planet earth) as they had envisioned.

One major spokespersons of a prominent Signore family spoke on MSNBC Business News to suggest a global tax perhaps. The question is: Pay a global tax to whom, G-20 capitalist or G-2 Capitalist? And if a global tax is collected, whom will regulate its usage?

New World Order-One World Government

New World Order (NOW) proposals led to the creation of international organizations, such as the United Nations and N.A.T.O., and international regimes,

---

[45] www.gata.org/files/GDBC_Report.pdf

such as the Bretton Woods system and the General Agreement on Tariffs and Trade, which were calculated both to maintain a balance of power as well as regularize cooperation between nations, in order to achieve a peaceful phase of capitalism.[46] Of course these proposals have failed miserably. Wars continue to rage.

President Woodrow Wilson, Winston Churchill and George W. Bush Sr. used the term "new world order" (NWO) during the 20th century to refer to a new dramatic change in the Caucasians rule over money, and the imposition of power to manage the nations of the earth, under the rule of white authority. The Caucasian concept of a new world order is an opportunity to implement global commercialization in a sense to open new business markets everywhere as long as Europe and America organize, supervise and deputize the ordeal.

To make westernized new world order dreams complete, every nation on earth would need to pay a global tax, which means every person within its realm of legal fiction will reclassify self as a legal person (sometimes referred to as a juristic person or a body corporate as person). Rather you know it or not, this is exactly how every U.S. citizen is recognized in North America.

In 1992, Dr John Coleman published conspirators' Hierarchy: The Story of the Committee of 300. On page 161 of the Conspirators Hierarchy, Dr Coleman wrote: "A

---

[46] en.wikipedia.org/wiki/New_World_Order_(conspiracy_theory)

One World Government and one-unit monetary system, under permanent non-elected hereditary oligarchists who self-select from among their numbers in the form of a feudal system as it was in the Middle Ages...There will be no middle class, only rulers and the servants. All laws will be uniform under a legal system of world courts practicing the same unified code of laws, backed up by a One World Government...The system will be on the basis of a welfare state; those who are obedient and subservient to the One World Government will be rewarded with the means to live; those who are rebellious will simple be starved to death or be declared outlaws..."

There is a wealthy dark side of the Caucasian race who are absolute non-believers in the laws of Allah with respect to prohibitions on profiteering with usurious loans. The history of such unjust profiteering goes back 650 years starting with the Lombard's of Italy and yet exist within the lending administrative apparatus of America and Europe's Signore Banking Families.

Earlier I wrote why Christian kings used the Jewish merchant class to issue loans, and why those Jews obliged. First, no king wanted to be excommunicated by the Holy See of Western Rome; second, Jewish merchants could be used as scapegoats, and third, Jews were making money.

According to JewishEncyclopedia.com vol. 12, *"The Church, basing itself upon a mistranslation of the text Luke vi. 35 interpreted by the Vulgate "Mutuum date, nihil inde sperantes," but really meaning "lend, never despairing" declared any extra return upon a loan as against the divine law, and this prevented any mercantile use of capital by pious Christians. As the canon law did not apply to Jews, these were not liable to the ecclesiastical punishments, which were placed upon usurers by the popes, Alexander III in 1179 having excommunicated all manifest usurers. Christian rulers gradually saw the advantage of having a class of men like the Jews who could supply capital for their use without being liable to excommunication, and the money trade of western Europe by this means fell into the hands of the Jews united and organized as a legal corporation while Christian kings lust for war and power eventually subjected them to debtor status under the legal Jurisdiction of the Central bank of Europe.*

"[In] 1745 encyclical of Benedict XIV, forbids all interest...The Church then 'flip-flops' in 1830 when the Holy Office, with the approval of Pius VIII, allows the justifiable taking of interest. Not only is the taking of interest now allowed, but the 1917 Code of Canon Law even said that religious orders were to keep their assets on deposit in interest bearing accounts."[47]

Holy See Accepted Usury

Usury represented free money to International merchants. When they came to realize the Holy See of

---

[47] http://frcoulter.com/presentations/usury/chapter1.html

Rome upheld its prohibition, they agreed that such a decree and office must lose its sovereignty within Western Christendom. That is cannon law interfered with mercantile use of capital therefore "pious Christians" were seen as enemies of Jewish lenders. As far as merchants and lenders were concerned, the demand for credit had to increase thus necessitating a modification in the definition of the term usury. Certain Christian governments began to make adjustments for usury. In 1545 England fixed a legal maximum interest, a practice later followed by other Western nations.[48]

Banks Rule The World

Overtime, *International Bankers* ultimately gained Juristic person statues, under Admiralty/maritime law to enforce commercial insurance that protected their investment dollars and products. This plan was set in place during the 16th century to guarantee both the king and the bank premiums are paid for certain, whereas the return is uncertain. In other words, any investment loses are passed onto the consumers via tax increases.

In Islam, if business is for government effects, insurance is lawful because the government's money is a collective of the people's money via taxes collected by the government. If the investment is a private affair, commercial insurance is unlawful. A private person's investment idea is an individual risk, not national risk,

---

[48] http://frcoulter.com/presentations/usury/chapter1.html

and the reward is an individual return for self and family first and foremost. In the case with Europe's Central Banks, they were and are privately owned. Therefore, all investment is a private risk! This is why what has been called "Bank Bail Outs" of the Bush and Obama Presidency era is one of the worst crimes millions of people have yet to comprehend.

The Church Is Conquered By International Bankers

The Church was relegated to function as an incorporate body combined into one body or unit; leading up to 1929 when the Holy See (seat of the pope) lost the war against the Central bank of England. His office was forced to focus on "its" own sovereignty and revenue flow when bank law overcame religious law. This action was realized in 1929 when *The Lateran Treaty of 1929* or Lateran Accords, between the Government of Italy and the Holy See, ratified the treaty on June 7, 1929. This treaty ended what was called the "Roman Question". These questions were in three documents:

1. A political treaty recognizing the full sovereignty of the Holy See in the State of Vatican City, which was thereby established. Prior to this treaty for 59 years, the Pope of Rome refused to leave the Vatican in order to avoid any appearance of accepting the authority wielded by the Italian government over Rome as a whole.

2. A concordat regulating the position of the Catholic Church and the Catholic religion in the Italian state in terms of tax

revenue, how might the church receive its tax payments, and the primacy of Roman Catholicism as the Italian state religion.

3. A financial convention agreed on as a definitive settlement of the claims of the Holy See following the losses of its territories and property between 1861 and 1929.

Today Vatican City is a city-state that came into existence in 1929 and the "The Institute for Works of Religion" commonly known as Vatican Bank is the Popes banks. In the beginning, the Italian government of Benito Mussolini, under the Lateran treaties of 1929 with the Holy See, paid a monthly salary to Catholic clergymen. This money was meant as a compensation for the nationalization of Church properties after the unification of Italy and Rome. This salary was called the congrua.

Then you have the "eight per thousand law" created as a result of an agreement, in 1984, between the Italian Government and the Holy See. Under this law Italian taxpayers are able to vote who shall receive 0.8% ('eight per thousand') of the total amount raised by income taxes.

Although the Holy See is closely associated with Vatican City, the Holy See is sovereign and operates independent within its territory. The two entities are separate and distinct. The term Holy See does not refer to Vatican City but to the Pope's spiritual and pastoral seat of governance he exercises through the Roman Curia i.e., administrative apparatus—or important members of the entire Catholic Church who justify group functions to achieve its goals.

A financial convention agreed on as a definitive settlement of the claims of the Holy See following the losses of its territories and property between 1861 and 1929. Again, the offices of Vatican City are not part of the Catholic Church. Vatican City is partially a commercialized religious zone on a 108.7-acre enclave of land wherein the Holy See acts as independent personality in international matters. Although this artificial state has been reduced in landmass, 172 states maintain diplomatic relations with the Holy See, and half of those that have accredited their ambassador to the Holy See.

The unique, noncommercial economy of Vatican City is supported financially by contributions (known as Peter's Pence) from Roman Catholics throughout the world, the sale of postage stamps and tourist mementos, fees for admission to museums, and the sale of publications. The Vatican has its own financial system and banks, with interests worldwide.[49]

Vatican City contains a population of approximately 3,000 people with less than 1,500 passport carrying "citizens". After completing their assigned appointments citizenship ceases.

In spite of all these fraudulent legal proceedings, Western Capitalism is ruined and International Central bank monetary policies have bankrupted nearly every

---

[49] http://en.wikipedia.org/wiki/Economy_of_Vatican_City

commercial nation on earth with debt financing. Can this be the reason, Vatican City and the Pope are seriously pondering Islamic banking?

For example, on Thursday, March 12, 2009 in the Brussells Journal, it said, *"the Vatican says Islamic finance system may help Western banks in crisis as alternative to capitalism...Then on March 4, 2009 Bloomberg Press wrote: "'Vatican said banks should look at the rules of Islamic finance to restore confidence amongst their clients at a time of global economic crisis. The ethical principles on which Islamic finance is based may bring banks closer to their clients and to the true spirit which should mark every financial service," the Vatican's official newspaper Osservatore Romano said in an article in its latest issue late yesterday.*

Chapter TEN

Moral Hazard

One of America's greatest Black scholars and educators, W. E. B. DuBOIS, said during his Commencement Address, Talladega College, June 5, 1944: *"We come therefore to the vast impasse of today: to the great question, What was the initial right and wrong of the original Jacobs and Esau and of their spiritual descendants the world over? We stand convinced today, at least those who remain sane, that lying and cheating and killing will build no world organization worth the building. We have got to stop making income by unholy methods; out of stealing the pittances of the poor and calling it insurance; out of seizing and monopolizing the natural*  *resources of the world and then making the world's poor pay exorbitant prices...Not only have we got to stop these practices, but we have got to stop lying about them and seeking to convince human beings that a civilization based upon the enslavement of the majority of men for the income of the smart minority, is the highest aim of man."*

Religious exegesis confirms that Jacob represents the Caucasoid race, particularly, their ruling leadership class. Esau, on the other hand, represents the Asiatic aboriginal nations of the earth whose ancient progenitors abdicated their birth right thousands of years ago to assure the white race would get their Time to rule.

For the record, Jacob's race were prewritten in biblical and Quranic text to have a position in which they'd rule or have sovereignty over the aboriginal nations for 6,000 years. This is the only reason why the God permitted their unethical global leadership. Universal time provided that a wicked rule according to the desires of the people would be allowed. The Holy Quran clarifies how nations governed by black people were destroyed in the holy land, for fraudulent practices against the people, long before the any Caucasian town or city was established.

> "Seest thou not how thy Lord dealt with the `Ad (people); Of the (city of) Iram, with lofty pillars; The like of which were not introduced in (all) the land? And with the Thamud (people), who cut out huge) rocks in the valley? And with Pharaoh, Lord of Stakes? (All) these transgressed beyond bounds heaped therein Mischief (on mischief). Therefore Ruins of ancient Thumud did thy Lord pour on them a scourge of diverse chastisement: For thy Lord is (As a Guardian) on a watch-tower." (Quran 89:1- 89:14)

This Quranic passage demonstrates how some of the ancient black governments on many occasions implemented their central authority tainted with mischief. When government structures centralize or federalize power to commit fraudulent practices due to greed, it always leads to divine judgment. Federal simply means: the sovereignty of a central authority of well-educated, privileged wealthy elitists who rule over the populous.

Rather than properly use the economic power God permits the elite and well educated to grasp, they verily reach a period to enrich themselves by directing economic activities and profits toward their ilk subsequently triggering other unnecessary societal problems resulting from financial violence. Today Western elites so-called well-educated have been even worse. It is said among them: "crime is not a problem; it's a system" because their investments are protected under a moral hazard law.

> "In economic theory, a moral hazard is a situation where a party will have a tendency to take risks because the costs that could result will not be felt by the party taking the risk. In other words, it is a tendency to be more willing to take a risk, knowing that the potential costs or burdens of taking such risk will be borne, in whole or in part, by others.

> "A moral hazard may occur where the actions of one party may change to the detriment of another after a financial transaction has taken place.

> "Moral hazard arises because an individual or institution does not take the full consequences and responsibilities of its actions, and therefore has a tendency to act less carefully than it otherwise would, leaving another party to hold some

responsibility for the consequences of those actions."[50]

Under western civilization governance, the peoples' taxes cover the losses created by "the elite decision-makers or Banking Families" who are protected by the institution of government. Simply put: moral hazard is a legal fiction invented by a criminal people for a criminal people.

> *"According to research by Dembe and Boden, the term dates back to the 1600s, and was widely used by English insurance companies by the late 1800s. Early usage of the term carried negative connotations, implying fraud or immoral behavior (usually on the part of an insured party). Dembe and Boden point out, however, that prominent mathematicians studying decision making in the 1700s used "moral" to mean "subjective", which may cloud the true ethical significance in the term."[51]*

The Quran says, *"Whole societies have passed away before your time, so travel about the earth and see the final fate of the deniers."* (Qur'an, 3:137)

As for the mischief of the Western world or Caucasian civilization, they are merely fulfilling a divine script rehearsed before they ever came into power and authority over the land. This script was recorded and rehearsed on the Arabian Peninsula and Asia Minor thousands of years ago both its successes and failures.

---

[50] http://en.wikipedia.org/wiki/Moral_hazard
[51] http://en.wikipedia.org/wiki/Moral_hazard#History_of_the_term

Chapter ELEVEN

United States vs. Nation Of Islam

America decided to declare the nature of her government under the sign of vulture or the Eagle, in 1787. One hundred and forty-three (143) years later, in 1930, the Son of Black Man came from Mecca, Arabia (East) to establish the Nation of Islam in North America. His name is Wallace Fard Muhammad. For teaching Islam to the black man and women in America, the United States Government had him arrested for the rulers of this land came to understand the Person before them.

> *"In the Holy City Mecca far away A Saviour was born...Some called Him Jesus, Some called Him Christ, But none call Him just right...Old Satan was slick Thought no one knew Where we would be found Nor what to do....But Allah came for He alone knew Giving Names Conquering old Satan too...Old got shaky When at last they found This man in jail Was God they had bound...They brought Him before justice And when He they had seen Ask his Authority Asked what did He mean?...He opened the Holy Quran So it could be seen Then boldly declared I AM THE SUPREME BEING!"[52]*

Around 1932 thereafter, the Nation of Islam (NOI) became a subject and target of FBI even up to this day, the FBI continues spying on the Nation of Islam to destroy its purpose.

---

[52] Written by Burnsteen Sharrieff. She was the Treasurer and Personal Secretary of Master W. Fard Muhammad during the early 1930's.

War against the Islamic people of the east began centuries ago and yet persists today. This war is being waged by the same tribes and people whom migrated out of the caves and hillsides of Europe over 4,000 years ago. What is the war about? It has always been about land, wealth ownership and divine recognition and the real known about the secret of Jesus (Isa).

Recall the Christian Crusades in 1095 A.D. led by Normans, French barons and Knights (sons of European kings). What was their goal? Answer: land, wealth and divine recognition and the real known about the secret of Jesus (Isa). To achieve these means also meant increased authority—the keys to the kingdom—planet earth. During the middle Ages, savage European Christians launched military campaigns to take the Holy Land from the Muslims, Christians and the Jews of Jerusalem.[53]

So why did the founding fathers of America decide to choose a bird of prey as the U.S. national symbol? What message did they want to convey?

"In May of 1782, the brother of a Philadelphia naturalist provided a drawing showing an eagle displayed as the symbol of "supreme power and authority." Congress liked the drawing, so before the end of 1782, an eagle holding a bundle of arrows in one talon and an olive branch in the other was accepted as the seal. The image was completed with a shield of red and white stripes covering the breast of the bird; a crest above the eagle's head, with a cluster of

---

[53] http://gbgm-umc.org/umw/bible/crusades.stm

thirteen stars surrounded by bright rays going out to a ring of clouds; and a banner, held by the eagle in its bill, bearing the words E pluribus unum. Yet it was not until 1787 that the American bald eagle was officially instituted."[54]

Those who study birds understand eagles and vultures are birds of prey that survive by plundering fragile or unsuspecting creatures and/or dead creatures. If these creatures are spiritually deaf, dumb, blind, ignorant and brainwashed human beings, U.S. government policy is designed to plunder your resources to survive another day.

Since 1930, U.S. government policy has been seeking to plunder the Nation of Islam (NOI) in the West. The question to consider is: What threat does the NOI represent to the U.S. government? What does the FBI know about W. Fard Muhammad and what he revealed to Elijah Muhammad and his people during the 1930's?

According many FBI reports, the NOI does not advocate violence. However, in one specific 1969 report, it does state the NOI must be disrupted at all cost. Read the following FBI report of 1969:

"The Nation of Islam & U.S. Government Counterintelligence Program (FBI File Date: 01-07-1969) SAC, Chicago (157-2209) January 7, 1969 Director, FBI (100-448006) COUNTERINTELLIGENCE PROGRAM BLACK NATIONALIST - HATE GROUPS RACIAL INTELLIGENCE (NATION OF ISLAM) strictly religious and self-improvement orientation, deleting the race hatred and separate nationhood aspects.

---

[54] www.baldeagleinfo.com/eagle/eagle9.html

"The alternative to **changing the philosophy of the NOI is the destruction of the organization**. This might be accomplished through generating factionalism among the contenders for Elijah Muhammad's leadership or through legal action in probate court on his death. Chicago should consider the question of how to **generate the factionalism necessary to destroy the NOI by splitting into several groups**. [BUREAU DELETION]

"In this connection Chicago should consider what counterintelligence action might be needed now or at the time of Elijah Muhammad's death to bring about such a change in NOI philosophy. Important considerations should include the identity, strengths, and weaknesses, of any contenders for NOI leadership. What are the positions of our [BUREAU DELETION] **informants** in regard to leadership? How could potential leaders be turned or neutralized?

"Although the Nation of Islam (NOI) does not presently advocate violence by its members, the group does preach hatred of the white race and racial separatism. The membership of the NOI is organized and poses a real racial threat. The NOI is responsible for the largest black nationalist newspaper, which has been used by other black extremists.

"The NOI appears to be the personal fiefdom of Elijah Muhammad. When he dies a power struggle can be expected and the NOI could change direction. We should be prepared for this eventuality. We should plan how to change the philosophy of the NOI to one of the strictly religious and self-improvement orientation, deleting the race hatred and separate nationhood aspects.

By all intend and purpose, the U.S. Federal government was and is at war against the Nation of Islam in North America?

After reading the next FBI letter, you'll see why FBI supervisors sent letters to special agents in charge of destroying the NOI. Their main target was and is the NOI Chicago Headquarters.

"The Nation of Islam & U.S. Government Counterintelligence Program Letter to Special Agent in Charge (SAC), Chicago RE: COUNTERINTELLIGENCE PROGRAM; (NATION OF ISLAM) 100-448006

"Legal action against the NOI on the death of its leader depends on the answers to several questions:

"In whose name are other NOI assets, such as mosque buildings, school buildings, the newspaper, Elijah Muhammad's homes, and NOI businesses?

"...Depending on the answers to these questions, probate law in Illinois, and whether Chicago might have a confidential source in probate administration, tying up the NOI in probate administration might be possible.

"Chicago should examine the NOI from the above counterintelligence angle and advise the Bureau. Consider the possibility of drawing up specific counterintelligence recommendations, to be acted upon when necessary, with various contingencies covered."

I reiterate, the FBI's hidden agenda was to disrupt the NOI's national interests as in financial matters— "black" economics and its truth about the real devil nature of the white race.

Although the latter aspect has been toned, the Nation of Islam is yet targeted by U.S. Homeland Security during these final hours of Judgment.

In 2010, according to news reports, Minister Louis Farrakhan and the Nation of Islam are under investigation by U.S. Homeland Security. The actual report on the Nation was not made public. Information was gathered from "open sources" beginning "around" February 2007, with agency officials violating protocols by taking longer than 180 days to determine whether the group and those being spied upon. So far we can see forces above the law—Shadow Government figures—are willing to break U.S. law to destroy the Nation of Islam i.e. Community of Jesus as it is cryptically mentioned in Holy Quran 3:54.

The actual documents to prove this point can be read on the Electronic Frontier Foundation website, which said; "While newspaper reports said Homeland Security improperly gathered information in 2007, as part of its speculation about The agency admitted that on October 12, 2007, the Department of Intelligence and Analysis (I&A) released a document titled "Nation of Islam: **"Uncertain Leadership Succession Poses Risk"**.

The document was disseminated via email to approximately 482 addresses, according to the agency. Included in that email listings were Homeland Security staff, representatives of federal departments and agencies, members of the intelligence community, educational institutions, law enforcement officials and congressional staff members, the agency said. "Legal experts, say there is no way of actually knowing whether the recipients of the intelligence information truly destroyed the info, or if anyone still has the documents in their possession.

What does Homeland Security hope to gain by

awaiting the departure of Minister Louis Farrakhan in terms of feeling that "Uncertain Leadership Succession Poses [a] Risk." What risk?

Why has the Federal government spend millions to hire African American spies to masquerade as NOI members with an intention to undermine the leadership and/or future leadership as far back as 1969 and even during the 1930's?

How far will the Federalized Powers go to prevent Black Islamic leadership from governing its own the Nation of Islam?

The greatest question of all is: Will there be an eminent attack upon the Nation of Islam to satisfy what the FBI and Homeland Security has been paid to do? Do Federalized Signore Powers want to destroy NOI leaders or members from the landscape of mainstream USA and world affairs for standing on the side of TRUTH and the incoming New Universal God of Peace?

Let's see: On June 9, 1996 at Mosque Maryam in Chicago Minister Farrakhan delivered a speech entitled "THE DIVINE DESTRUICTION OF AMERICA: Can She Avert It? In part, he announced this warning:

"The Honorable Elijah Muhammad said America is a preserved area. No foreign nation will destroy America lest they get the credit. God will do it Himself. He said America is number one on God's list to be destroyed. He talked about the four beasts: America, England, Germany and Italy. These are four white nations that have acted with great wickedness

against the darker people of the world. He says the more you fight against Elijah, the more God makes perfect through me. Elijah Muhammad said, we will make the truth so plain that a fool will find it hard to err.

"He said that he wants every no good disbeliever off the face of the earth. He said after this, America will never rise to be an independent nation again. She won't have the power that she had last year, meaning that every year from then her power would be diminished. He said America is gone, her money is gone. There is no more silver, no more gold, after a while she may ask for copper. Right now the money has no value. This (dollar) used to be a silver certificate that said pay the bearer on demand one dollar in silver. You can't get silver for it anymore; you can't get gold for it; it is the wealth of the country (gross national product) that is holding up the dollar. Elijah Muhammad said soon you will see money piled and burning in the streets. There will be another medium of exchange not coming from this country.

"He said America will continue to fight. He mentioned Vietnam and other areas of Asia. But the last fight, he said, will be in East Germany, and there she (America) will be destroyed.

"You think East Germany and West Germany have united. They broke the wall down but they are not united East Germany has been 40 years under the communists and West Germany has been under American and Western democracy. These are like apples and oranges, and it is very difficult to produce the integration. When the power of the Soviet Union was broken, many of you that were in the armed forces in Europe were called back to America. You don't have jobs but soon they are going to ask you to go back to Europe. The Messenger said that you would go away in the thousands and you will return in the tens. And when you return there will be no such thing as an American government.

"He said that all of America's possessions will be taken one by one, and all of her fortifications will be destroyed. Her great armies poured into Asia and Europe will not return. He said France will emerge, that France won't be with England and America. He said that England is going to trick America into war. He said white America will be destroyed mercilessly, that not even her own (white) brother or sister will show mercy on her. He said, it's not 40 years away, it's not 40 months away. He said preparation is being made to take her right now.

"The war is building and the four great judgments that God is going to use are rain, hail, snow and earthquakes. If you look at the weather, you already know that something is going on. When it rains, it's unusual rain that comes down in torrents. It is interfering with the foundation of the houses, tearing up the streets and the roads, upsetting the railroad tracks.

"The final thing is the destruction. The Honorable Elijah Muhammad told us of a giant Mother plane that is made like the universe, spheres within spheres. White people call them unidentified flying objects (UFOs). Ezekiel, in the Old Testament, saw a wheel that looked like a cloud by day but a pillar of fire by night. The Hon. Elijah Muhammad said that that wheel was built on the island of Nippon, which is now called Japan, by some of the original scientists. It took [150 million dollars in gold] at that time to build it. It is made of the toughest steel. America does not yet know the composition of the steel used to make an instrument like it. It is a circular plane, and the Bible says that it never makes turns. Because of its circular nature it can stop and travel in all directions at speeds of thousands of miles per hour. He said there are 1,500 small wheels in this mother wheel, which is a half-mile by a half-mile. This Mother Wheel is like a small human built planet. Each one of these small planes carry three bombs.

"The Honorable Elijah Muhammad said these planes were used to set up mountains on the earth. The Qur'an says it like this: We have raised mountains on the earth lest it convulse with you. How do you raise a mountain, and what is the purpose of a mountain? Have you ever tried to balance a tire? You use weights to keep the tire balanced. That's how the earth is balanced, with mountain ranges. The Honorable Elijah Muhammad said that we have a type of bomb that, when it strikes the earth a drill on it is timed to go into the earth and explode at the height that you wish the mountain to be. If you wish to take the mountain up a mile, you time the drill to go a mile in and then explode. The bombs these planes have are timed to go one mile down and bring up a mountain one mile high, but it will destroy everything within a 50 square mile radius. The white man writes in his above top-secret memos of the UFOs. He sees them around his military *installation like they are spying.*

"That Mother Wheel is a dreadful looking thing. White folks are making movies now to make these planes look like fiction, but it is based on something real. The Honorable Elijah Muhammad said that Mother Plane is so powerful that with sound reverberating in the atmosphere, just with a sound, she can crumble buildings. And the final act of destruction will be that Allah will make a wall out of the atmosphere over and around North America. You will see it, but you won't be able to penetrate it. He said Allah (God) will cut a shortage in gravity and a fire will start from 13-layers up and burn down, burning the atmosphere. When it gets to the earth, it will burn everything. It will burn for 310 years and take 690 years to cool off.

"The Book of Revelation says, And the Kings of the earth who have committed fornication with her, shall lament for her when they shall see the smoke of her burning. This fire is for us. It's prepared from men and stones. Stones represent the hard-hearted people of this wicked world and for men who refuse to change and come to God.

"You are in the valley of decision. What are you going to do? Are you going to clean up your lives? I'm not asking you if you want to join me. You can if you want to. But if you are in the church, you better make the church right because Judgment is going to begin at the so-called house of God. Wherever you are, you are going to have to clean it up. Whatever we are doing that we know is wrong, we must straighten it out. But if

you don't it's on you.

"I hope and pray that I have made the message clear. Thank you for reading these few words."

"As Salaam Alaikum
Honorable Minister Louis Farrakhan Muhammad"

Chapter TWELVE

The Mother Plane By Elijah Muhammad

The Central Intelligence Agency (CIA) is an independent US Government agency responsible for providing national security intelligence to senior US policymakers. In late 1993, after being pressured by UFOlogists for the release of additional CIA information on UFOs, DCI R. James Woolsey ordered another review of all Agency files on UFOs. Using CIA records compiled from that review, this study traces CIA interest and involvement in the UFO controversy from the late 1940s to 1990. It chronologically examines the Agency's efforts to solve the mystery of UFOs, its programs that had an impact on UFO sightings, and its attempts to conceal CIA involvement in the entire UFO issue. What emerges from this examination is that, while Agency concern over UFOs was substantial until the early 1950s, CIA has since paid only limited and peripheral attention to the phenomena.[55]

As early as 1931, the Honorable Elijah Muhammad was taught about a dreadful plane--UFO in the sky by the Supreme Being, Master W. Fard Muhammad. The *Mother* of those UFO's was identified as the Mother Plane designed to carry 1,500 smaller crafts—ufo's. The architect of this Plane is none other than the Divine

---

[55] https://www.cia.gov/library/center-for-the-study-of-intelligence/csi-publications/csi-studies/studies/97unclass/ufo.html

Supreme Being. Therefore, to members of the Nation of Islam of the West, the knowledge about "UFO's" is common knowledge. The Honorable Elijah Muhammad said:

"Today, that Plane is in the air and the scientists and the astronomers of America, have seen it here. Mr. Khowule, an ex-General of the United States Armed Forces of America, bears me witness that it is up there, and have dotted around America's planes in the sky, here in America. He wrote a book on it and condemns America for keeping it a secret.[56] But the Government wants to make it all a lie, that there is nothing like this going on, that there is no truth to no flying saucers! Maybe there is no truth to no 'NATURAL FLYING SAUCER," but there is plenty of truth to CIRCULAR PLANES!!! Planes made like a wheel up in the sky!!! There's plenty of truth to that, and the scientists and astronomers bear me witness here in America, that it's up there.

"The Mother Plane was shown to me, now 30 years ago, [in 1931] by God Himself (Master Fard Muhammad). I have seen this Plane and have sketched it many times. As Almighty God (Allah) sketched it to me, I followed the sketching of His. Today, you would like to dismiss it as a lie, but those who peep through the glass (the telescope) and see this plane, do not consider it a lie.

"...An absolute Man-Made-Planet, moving at such terrific

---

[56] http://www.davidicke.com/forum/archive/index.php/t-95087.html

speed that you could hardly imagine it; flickering amid the stars of Heaven. Out of the reaches of your (the Devil's) fire, you can't reach it, you can't shoot it down; it's impossible! You can't go to it, you can't capture it; it's impossible...

"Hence, there is Zachariah, Chapter 5:1- 2, we are told of a "Flying Roll" which is THIRTY FEET long and FIFTEEN FEET wide. This is a near perfect description of one of Allah's Circular Planes. Again, the key words 'flying roll' appear which is an adequate description for the scout ships (Circular Planes) that are aboard the Mother Ship. The ignorant will say that the second verse should be taken on face value. But this is impossible, since the  *measurements of this 'flying roll' is given by the Prophet. . . the length thereof is twenty cubits.' Note: one cubit is an ancient linear measure of about 18 inches, according to the dictionary. Since one cubit is 13 inches, twenty cubits equal 360 inches, or thirty feet, which is the length of this 'Flying Roll' ...and the breadth thereof is ten cubits; ends the second verse. Ten cubits equal 180 inches (15 feet) which is the width of this 'flying roll'...*[57]

The photo above is a "flying roll" or scout ship. It was not made by the British nor American governments; nor the Russian, German, Chinese or Japanese

---

[57] The Fall of America CHAPTER 58, The Mother Plane, by Elijah Muhammad

Governments. These type of planes were made by the original Asiatic Scientists (Angels) of the planet earth under the direction of Master W. Fard Muhammad—the Divine Supreme Being who some religious writers described as the Son of Man. UFO's are His weapons made for the Final Battle and made to bring about the New Universal Government. (Matthew 24:30)

In 2009, the Mother Plan gave a demonstration over Mexico. After reviewing video footage provided on Youtube (Ufo Mothership Over Mexico City Interview 22 / may / 09), there should be no doubt about the truth revealed by the Honorable Elijah Muhammad and Minister Louis Farrakhan Muhammad.

"The Mother Plane was made to destroy this world of evil and to show the wisdom and mighty power of the God Whom came to destroy an old world and set up a new world. The nature of the new world is righteousness. The nature of the new world cannot be righteousness, as long as unrighteousness is in its midst. The same type of plane was used by the Original God to put mountains on His planets.

"Allah (God) Who came in the Person of Master Fard Muhammad, to Whom praises are due forever, is wiser than any god before Him as the Bible and the Holy Qur'an teach us. He taught me that this place will be used to raise mountains on this planet (earth). The mountains that He will put on this earth will not be very high. He will raise these mountains to a height of one (1) mile over the United States of America.

There are planes in various nations today, but this is the mother of them all. Why? Because this type of plane was used before the making of this world. Why should God make such a sign of His power to destroy a nation? Because this is the final destruction of that people who have opposed God in His purpose and aims for Justice and Righteousness.

"The Mother Plane, according to what has been described of it by the devil scientists, is capable of not only staying up for long periods of time; but it is also capable of eluding the scientists. They want to attack and destroy it; but if a plane did get close enough to attempt to carry out this purpose, it would be destroyed instead. The white man has learned that this is not a place to be played with. Planes come out of the Mother Plane.

"In the 1930's Canadian newspapers reported that they saw the wheel (Mother Plane). It came down out of the sky. They admitted that it looked like a great city, and that something came down from it; it appeared to be a tube, but the tube-like thing went back up again. Allah (God) Who came in the Person of Master Fard Muhammad, to Whom praises are due forever, taught me that after six months to a year, the Mother Plane comes into the gravity of the earth. It takes on oxygen and hydrogen in order to permit it to stay out of the earth's gravity until it needs refueling again."

The Honorable Elijah Muhammad further taught:

"Ezekiel saw the Mother Plane in a vision. According to the Bible, he looked up and saw this Plane (Ez. 1:16) and he called it a wheel because it was made like a wheel. A Plane that is wheel-shaped can turn in any direction, at any time. He admitted that the Plane was so high that it looked dreadful, and he cried out, "O wheel" (Ez. 10:13).

Ezekiel saw great work going on in the wheel and four living creatures 'and their work was as it were a wheel in the middle of a wheel.' (Ez. 1:16). And when the living creatures went, the wheels went with them: and when the living creatures were lifted up from the earth, the wheels, were lifted up Ez. 1:19. The power of the lifting up of the four creatures was in the wheel. The four creatures represent the four colors of the original people of the earth. There are five great powers of the nations of the never thought the day would come when you would see your wife, mother and old grey-haired women walking down the street today, half-nude. A few years ago they would not have dared to come out into the public like that. But now they do so because the devil has put his approval on this kind of attire. They desire to please the devil. They do whatever the devil bids them to do. The devil desires to take the Black man with him to his doom...

"There is no known equal of the Mother Plane. This is the reason why she is called the Mother Plane. The Mother Plane is made for the purpose of destroying the present world. She has no equal. Do not marvel at the make of this plane, since it is from the God Who made the universe of floating planets and stars, which are supported only by the Power of Allah in their rotation in their orbits. Allah (God) Whom came in the Person of Master Fard Muhammad, to Whom praises are due *forever, taught me that the Mother plane is a little human-made planet...*

"The Mother Plane is capable of staying out of the earth's gravity for a whole year. He is capable of producing her own sphere of oxygen and hydrogen, as any other planet is able to do. The Mother Plane carries the same type of bomb on her that our Black scientists dropped on the planet earth to bring up mountains out of the earth after the planet earth was created. The knowledge of how to do this has not been given to the world (white race), nor will they ever get this kind of knowledge. The knowledge of the world is limited. If the devil would get this type of knowledge we could just say that we are goners. However, they are not able to attain this type of knowledge...The knowledge and power of this world's life (white race) is limited. The world of the white man was made from what he found and what he has seen and learned from the work of the original Black man. The white race is far from being able to equal the power and wisdom of the original Black man.

"The Mother Plane and her work is a display of the power of the mightiest God, Master Fard Muhammad, to Whom praises are due forever. Master Fard Muhammad, to Whom praises are due forever, is the Wisest and Best Knower; He is the Mightiest of Them All...

"In Ezekiel's vision concerning the wheel, he said that he heard the voice of one tell the other to take coals of fire and to scatter it over the cities; this means bombs. It could mean fire too, however. The Plane is to drop bombs, which would automatically be timed to burrow quickly to a position of one mile below the surface of the earth where they are timed to explode. Allah (God) taught me that these bombs are not to be dropped into water. They are to be dropped only on the cities. It will be the work of the wheel. The wheel is the power of the four creatures, namely the four colors of the Black man (Black, brown, yellow and red). The Red Indian is to benefit also from the judgment of the world...

"It is useless to try to ignore Ezekiel's vision of the wheel, for the make and the destructive work of the wheel was foretold before it came to pass. The disbeliever believes that which he sees present and not that which is prophesied to come. That is why he is the loser and takes the course to hell, because he disbelieves in that which is prophesied to come about a particular day. This is what the enemy is trying to do today with the Black Man. He is fascinating him with sport and play and indecency and the doing of evil to keep him from going to the God Who is present.

"O wheel," says the prophet Ezekiel. She was so high up in the sky that she looked dreadful. She is capable of staying away from you who plan the destruction of her. She is capable of confusing you who would try to reach her with your means of destruction. There are scientists on the Mother Plane who know what you are thinking about before the thought materializes (Holy Qur'an Ch. 50:16). Therefore, it is impossible to try to attack the Mother Plane. She can attack you, but you cannot attack her. The Mother Plane can hide *behind other stars and make herself invisible to the eye because she does not have to wait on a power from the earth. She can produce her own power to go wherever she desires to go in space."*

In 1956, USAF Capt. Edward J. Ruppelt, USAF Capt acknowledged "...Of these (UFO) reports, the radar-visual sightings are the most convincing. When a ground radar picks up a UFO target and a ground observer sees a light where the radar target is located, then a jet interceptor is scrambled to intercept the UFO and the pilot also sees the light and gets a radar lock only to have the UFO almost impudently out distance him, there is no simple answer..."

This is why the Honorable Elijah Muhammad gave unadulterated insight respecting the mission of the Mother Plane and its 1500 circular smaller planes to which the United States Airspace Dynamic Engineers cannot comprehend nor defeat in the final battle to establish the Kingdom of Heaven on Earth. So he further goes on the reveal:

"The Mother Plane is not like your little bullets or cameras which are powered by your limited power. The Bible prophecies that today Allah (God) wishes to make known to use that He is God. He wishes to be respected as the Superior God. He wishes all life in the universe to know that He is the Greatest. The Muslim recognizes Allah (God) to be the greatest. He always repeats 'Thou art the greatest. There is

no god Like unto Thee, None deserves to be served or worshipped besides thee.' "O mighty wheel. I repeat, that there is plenty of significance to the make of the Mother Plane. There is much significance to the course of operation of her work.[58]

The Image below was drawn by a Japanese Pilot during the 1990's while flying over Alaska. Notice the size of his airliner lower right side of the Mother Plane. Did not the Honorable Elijah Muhammad teach the Mother Plane is oval and 1 half mile by 1 half mile?

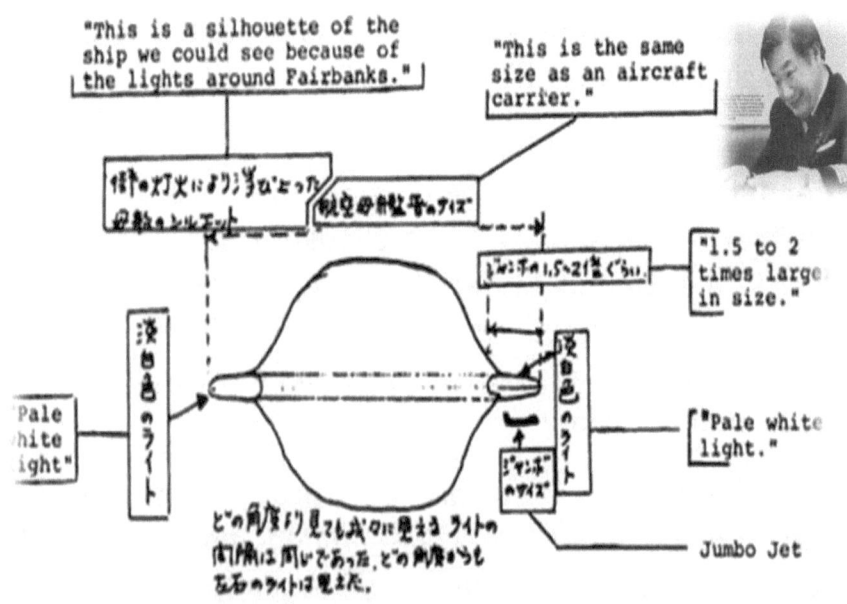

Figure 8: Capt. Terauchi's drawing, a month and a half after sighting, of "gigantic spaceship"

---

[58] The Fall of America CHAPTER 58, The Mother Plane, by Elijah Muhammad

Further evidence as to the size and shape of the Mother Plane is shown below in a picture taken by Captain Viegas in 1958 over Trindade Island Brazil.

According to Captain Viegas, the object was like a flattened sphere encircled at the equator by a large ring or platform. In Barauna's words, *"...it made no noise, although with the shouting of the people on the deck and the noise of the sea, I cannot be certain. It had a metallic look, of an ash color, and has like a condensation of a green vapor around the perimeter, particularly in the advancing edge."*

The Judgment of this current 6,000 year old world order was not left up to chance. It is based upon the whole life of Time given a criminal people, including some of the black, brown, red and yellow nations (who love the influence and socialization under "white authority"). However, the most innocent of peoples' under white authority is America's former slave descendants or legal captives—the only non-immigrant living in the hells of North America.

> "The original black man has been without the knowledge of himself for a long time and this one (the American so-called Negroes), of all of his kind, is the dumbest to the knowledge of self, due to the way his slave-master teaches and trains him. But this is the time of the awakening of this poor slave and no powers on earth or in the heavens above will be able to prevent it. For it is the will and work of Allah and His choice of the people. He has chosen the so-called Negroes, but they being blinded and made deaf and dumb, have not but a few chosen Allah to be their God; but they will, after they see more of His power displayed in the West -- and they will see it. It is going on now. It is a must with Allah to restore the lost sheep.

> "The black people are by nature the righteous. They have love and mercy in their hearts even after trying to live the life of the devils - - this is still recognized in them. When they are fully in the knowledge of self, they will do righteousness and live in peace among themselves. One can't judge

them now for they are not their own selves. We, the original nation of earth, says Allah, the Maker of everything -- sun, moon and stars and the race called white race -- are the writers of the Bible and Qur-an."[59]

To conclude this book, let us pray there is no U.S. government attack on the Nation of Islam before or after the departure of the Honorable Minister Louis Farrakhan in the name of a war against Islam. The assignment he and those with him carry represents the plan of God Himself.

Just as the Caucasian needed Gold to get thier economic wheels turning, so does Black America—the lost and found members of the tribe of Shabazz. The shadow government must remember the bible passage that reads, *"The kings of Tarshish and of distant shores will bring tribute to him; the kings of Sheba and Seba will present him gifts."* (Psalm 72:10)

According to the teachings of the Honorable Elijah Muhammad, a deal will be worked out between America's wise Moslem Sons and the Nation of Islam. But what will compel such a deal? Is this deal related to the work of the Mother Plane? Essentially, this worlds ruling Signore families are waiting to first see the wrath of Allah in what he has created as a weapon—called the Mother Plane or Mother Plane or Wheel. Jewish Rabbi's are very well versed about the work of this great

---

[59] www.muhammadspeaks.com/Makingofdevil.html

mechanical space craft and weapon of God. They called it Merkabah thousands of years ago.

> "Merkabah/Merkavah mysticism (or Chariot mysticism) is a school of early Jewish mysticism, c. 100 BCE – 1000 CE, centered on visions such as those found in the Book of Ezekiel chapter 1, or in the hekhalot ("palaces") literature, concerning stories of ascents to the heavenly palaces and the Throne of God.

> "The main corpus of the Merkabah literature was composed in Israel in the period 200–700 CE, although later references to the Chariot tradition can also be found in the literature of the Chassidei Ashkenaz in the Middle Ages. A major text in this tradition is the Maaseh Merkabah (Works of the Chariot).[60]

So what will it take before this worlds evil rulers (Signore Banking Families or Shadow Government) BOW DOWN and through down their crowns at the feet of the NEW RULERS?

60

Appendix 2

List of IMF Member Countries

The International Monetary Fund (IMF) is an organization of 188 countries, working to foster global monetary cooperation, secure financial stability, facilitate international trade, promote high employment and sustainable economic growth, and reduce poverty around the world.

Membership of the **IMF** (Date of entry into force: December 27, 1945) Chronological List (188 Member Countries)	
Member	**Effective Date of Membership**
Belgium[1]	December 27, 1945
Bolivia[1]	December 27, 1945
Canada[1]	December 27, 1945
China[1]	December 27, 1945
Colombia[1]	December 27, 1945
(Czechoslovakia)[1,2,3]	(December 27, 1945)
Egypt[1]	December 27, 1945
Ethiopia[1]	December 27, 1945
France[1]	December 27, 1945
Greece[1]	December 27, 1945
Honduras[1]	December 27, 1945
Iceland[1]	December 27, 1945
India[1]	December 27, 1945

Iraq[1]	December 27, 1945
Luxembourg[1]	December 27, 1945
Netherlands[1]	December 27, 1945
Norway[1]	December 27, 1945
Philippines[1]	December 27, 1945
South Africa[1]	December 27, 1945
United Kingdom[1]	December 27, 1945
United States[1]	December 27, 1945
(Yugoslavia)[1,4,5]	(December 27, 1945)
Dominican Republic[1]	December 28, 1945
Ecuador[1]	December 28, 1945
Guatemala[1]	December 28, 1945
Paraguay[1]	December 28, 1945
Iran, Islamic Republic of (Iran)[1]	December 29, 1945
Chile[1]	December 31, 1945
Mexico[1]	December 31, 1945
Peru[1]	December 31, 1945
Costa Rica[1]	January 8, 1946
(Poland)[1,6]	(January 10, 1946)
Brazil[1]	January 14, 1946
Uruguay[1]	March 11, 1946
(Cuba)[1,7]	(March 14, 1946)
El Salvador[8]	March 14, 1946
Nicaragua[8]	March 14, 1946
Panama[8]	March 14, 1946
Denmark[8]	March 30, 1946

Venezuela, República Bolivariana de[8]	December 30, 1946
Turkey	March 11, 1947
Italy	March 27, 1947
Syrian Arab Republic (Syria)	April 10, 1947
Lebanon	April 14, 1947
Australia	August 5, 1947
Finland	January 14, 1948
Austria	August 27, 1948
Thailand (Siam)	May 3, 1949
Pakistan	July 11, 1950
Sri Lanka (Ceylon)	August 29, 1950
Sweden	August 31, 1951
Myanmar (Burma)	January 3, 1952
Japan	August 13, 1952
Germany	August 14, 1952
Jordan	August 29, 1952
Haiti	September 8, 1953
(Indonesia)[9]	(April 15, 1954)
Israel	July 12, 1954
Afghanistan, Islamic Rep. of (Afghanistan)	July 14, 1955
Korea	August 26, 1955
Argentina	September 20, 1956
Vietnam (Viet Nam)	September 21, 1956
Ireland	August 8, 1957
Saudi Arabia	August 26, 1957

Sudan	September 5, 1957
Ghana	September 20, 1957
Malaysia (Malaya)	March 7, 1958
Tunisia	April 14, 1958
Morocco	April 25, 1958
Spain	September 15, 1958
Libya	September 17, 1958
Portugal	March 29, 1961
Nigeria	March 30, 1961
Lao People's Democratic Republic (Laos)	July 5, 1961
New Zealand	August 31, 1961
Nepal	September 6, 1961
Cyprus	December 21, 1961
Liberia	March 28, 1962
Togo	August 1, 1962
Senegal	August 31, 1962
Somalia	August 31, 1962
Sierra Leone	September 10, 1962
Tanzania (Tanganyika)	September 10, 1962
Kuwait	September 13, 1962
Jamaica	February 21, 1963
Côte d'Ivoire (Ivory Coast)	March 11, 1963
Niger	April 24, 1963
Burkina Faso (Upper Volta)	May 2, 1963
Cameroon	July 10, 1963

Central African Republic	July 10, 1963
Chad	July 10, 1963
Congo, Republic of	July 10, 1963
Benin (Dahomey)	July 10, 1963
Gabon	September 10, 1963
Mauritania	September 10, 1963
Trinidad and Tobago	September 16, 1963
Madagascar (Malagasy Republic)	September 25, 1963
Algeria	September 26, 1963
Mali	September 27, 1963
Uganda	September 27, 1963
Burundi	September 28, 1963
Congo, Democratic Republic of the (Zaïre)	September 28, 1963
Guinea	September 28, 1963
Rwanda	September 30, 1963
Kenya	February 3, 1964
Malawi	July 19, 1965
Zambia	September 23, 1965
Singapore	August 3, 1966
Guyana	September 26, 1966
Indonesia[9]	February 21, 1967
Gambia, The	September 21, 1967
Botswana	July 24, 1968
Lesotho	July 25, 1968
Malta	September 11, 1968

Mauritius	September 23, 1968
Swaziland	September 22, 1969
(Yemen, People's Democratic Republic of (Southern Yemen))[10]	(September 29, 1969)
Equatorial Guinea	December 22, 1969
Cambodia	December 31, 1969
(Yemen Arab Republic)[10]	(May 22, 1970)
Barbados	December 29, 1970
Fiji	May 28, 1971
Oman	December 23, 1971
Samoa (Western Samoa)	December 28, 1971
Bangladesh	August 17, 1972
Bahrain	September 7, 1972
Qatar	September 8, 1972
United Arab Emirates	September 22, 1972
Romania	December 15, 1972
Bahamas, The	August 21, 1973
Grenada	August 27, 1975
Papua New Guinea	October 9, 1975
Comoros	September 21, 1976
Guinea-Bissau	March 24, 1977
Seychelles	June 30, 1977
São Tomé and Príncipe	September 30, 1977
Maldives	January 13, 1978
Suriname	April 27, 1978
Solomon Islands	September 22, 1978

Cape Verde	November 20, 1978
Dominica	December 12, 1978
Djibouti	December 29, 1978
St. Lucia	November 15, 1979
St. Vincent and the Grenadines	December 28, 1979
Zimbabwe	September 29, 1980
Bhutan	September 28, 1981
Vanuatu	September 28, 1981
Antigua and Barbuda	February 25, 1982
Belize	March 16, 1982
Hungary	May 6, 1982
St. Kitts and Nevis	August 15, 1984
Mozambique	September 24, 1984
Tonga	September 13, 1985
Kiribati	June 3, 1986
Poland[1,6]	June 12, 1986
Angola	September 19, 1989
Yemen, Republic of[10]	May 22, 1990 [7]
(Czechoslovakia)[1,2,3]	(September 20, 1990)
Bulgaria	September 25, 1990
Namibia	September 25, 1990
Mongolia	February 14, 1991
Albania	October 15, 1991
Lithuania	April 29, 1992
Georgia	May 5, 1992
Kyrgyz Republic (Kyrgyzstan)	May 8, 1992

Latvia	May 19, 1992
Marshall Islands	May 21, 1992
Estonia	May 26, 1992
Armenia	May 28, 1992
Switzerland	May 29, 1992
Russian Federation	June 1, 1992
Belarus	July 10, 1992
Kazakhstan	July 15, 1992
Moldova	August 12, 1992
Ukraine	September 3, 1992
Azerbaijan	September 18, 1992
Uzbekistan	September 21, 1992
Turkmenistan	September 22, 1992
San Marino	September 23, 1992
Bosnia and Herzegovina[5]	December 14, 1992
Croatia[5]	December 14, 1992
Macedonia, former Yugoslav Republic of[5]	December 14, 1992
Slovenia[5]	December 14, 1992
Serbia[5]	December 14, 1992
Czech Republic[3]	January 1, 1993
Slovak Republic[3]	January 1, 1993
Tajikistan	April 27, 1993
Micronesia, Federated States of	June 24, 1993
Eritrea	July 6, 1994
Brunei Darussalam	October 10, 1995

Palau	December 16, 1997
Timor-Leste (East Timor)	July 23, 2002
Montenegro[5]	January 18, 2007
Kosovo	June 29, 2009
Tuvalu	June 24, 2010
South Sudan	April 18, 2012

Appendix 2

London Gold Fix: 1660-2004

1660 Gold price: £4.05 per troy/oz fine.

1661 The East India Company secured exclusive trading rights to the east. In next 45 years they shipped almost 500,000 troy/oz 15.5 m.t from London to India. 1663 The guinea, named after Guinea on Africa's 'gold coast', was first struck.

1671 Moses Mocatta set up in London, founding the firm that later became Mocatta & Goldsmid, the oldest members of the market. Nine generations of the family worked in the bullion market.

1676 Mocatta first sent gold to India via the East India Company.

1779 Abraham Mocatta took Asher Goldsmid as his partner. Lowndes London Directory recorded: Mocatta & Goldsmid (Brokers), Grigsby's Coffee House in 1783.

1785 Bank of England's Warehouse changed its name to The Bullion Office.

1789 Bank of England opened an account for Louise d'Or coins brought by French refugees escaping the revolution.

1797 Bank of England's gold reserve, drained by costs of the Napoleonic wars, was down to 235,000 troy/oz (£1 million) against note issue liabilities of £15.5 million. Cash payment in gold against bank notes was suspended on 20 February, and did not resume for twenty-four years.

1805 Nathan Mayer Rothschild opened his banking house in London and became closely involved in secret shipments of gold and silver to the Duke of Wellington's army in Europe against Napoleon. Mocatta & Goldsmid rounded up the gold, often bidding over the market price.

1810 The House of Commons Select Committee on the High Price of Bullion, which had risen from the normal £3.89 for 916 gold to £4.50. The evidence provided a unique insight into the London market, as Aaron Asher Goldsmid, Nathan Mayer Rothschild (incognito as a 'continental merchant') and gold refiner William Merle explained the trade. The Committee concluded the Bank of England had been printing too many notes as they were no longer redeemable in gold.

1811 Sharp & Kirkup, auctioneers since 1796, started brokerage in gold and silver, but refused Bank of England accreditation.

1815 The gold price jumped to £5.35 for standard gold after Napoleon escaped from Elba, but after his defeat at Waterloo fell back under £4.00.

1816 The Coinage Act made the gold standard official, with the guinea replaced by the sovereign, worth £1.00, weighing 0.25 troy/oz/7.77 g at 916 fine. The first sovereigns were issued in 1817.

1821 Full resumption of cash payments in gold against notes by the Bank of England.

1840 Bank of England's Bullion Office opened to

'any sworn broker', because of the increase in gold from Russia entering the Port of London, thus ending Mocatta's exclusive arrangement.

1848 California gold rush brought a new dimension to the gold market, with tripling of mine output by 1850.

1851 Australian gold discoveries in New South Wales and Victoria pushed world output to 6.5 million

troy/oz/203 m.t by 1855. Most Australian gold came to London, transforming the market.

1852 Stewart Pixley set up as a bullion broker, the first of four generations in the market, with William Haggard as partner. The firm later became Pixley & Abell.

1853 Samuel Montagu founded his bullion and exchange business (today part of HSBC).

1855 London market comprised: Brokers: Mocatta & Goldsmid, Sharps & Wilkins, Pixley & Haggard (shortly

Abell), Samuel Montagu & Co. 1856 Approved refiners: Johnson & Matthey, Browne & Wingrove, Rothschild's Royal Mint Refinery, H. L. Raphael's Refinery 'Good delivery' bars were of 200 troy/oz, and the Bank required a triple assay of each bar. After 1871, 400 t. oz bars were also accepted.

1871 Germany went on the gold standard; most other European nations followed suit.

1876 House of Commons Select Committee on Depreciation of Silver took expert advice from Mocatta, Pixley and Sharps on gold output coin fabrication and central bank stocks.

1886 Royal Commission on Gold & Silver investigated changed relationship between the metals. Samuel Montagu sat on the Commission, Stewart Pixley and Sir Hector Hay of Mocatta gave statistical briefings. The Commission came out in favour of the gold standard, as opposed to bimetallism.

1886 Gold discoveries in the Witwatersrand in South Africa. Output reached 3.8 million troy/oz/118.2 m.t by 1898. It came to London for refining and sale.

1893 Gold rush in Western Australia after discoveries at Kalgoorlie.

1896 US presidential election had bimetallism as the key issue, supported by William Jennings Bryan. He was defeated and the US went on the gold standard in 1900.

1897/9 Peak years of gold rush to the Yukon in Canada, yielding 3.7 million troy/oz/115 m.t.

1900 Gold price: £4.25 t. oz fine/US $20.67.

1914 At outbreak of World War I governments limited gold flows and called in much domestic coin, especially in Britain and France. The gold standard was never officially suspended, although in practical terms it was.

1919 The Bank of England, determined to restore London as the main gold market, reached an agreement with the seven South African mining houses to ship their gold to London for refining, after which it would be sold through N. M. Rothschild 'at the best price obtainable, giving the London market and the Bullion Brokers a chance to bid'. Thus, on 12 September 1919, the first gold fixing took place; the price was fixed at £4.94 (US $20.67) per troy/oz fine — a change from the previous price for standard 916 gold. The bids were made by telephone for the first few days and it was then decided to hold a formal meeting at Rothschild's offices in New Court, St Swithins Lane.

1922 The Rand Refinery was opened in South Africa, but the gold continued to be sold through London.

1925 Britain went onto a gold bullion standard at the old fixed rate of £4.25 t. oz fine, but with minimum purchase of 400 ounces.

1931 Britain and many other nations came off the gold standard, with the onset of the depression. Sterling was devalued creating price between £5.50 and £6.34.

1933 The US came off the gold standard. President Roosevelt stopped the convertibility of dollars into gold and ordered US citizens to hand in coin.

1934 On 31 January Roosevelt set a new fixed price of $35 per troy/oz. The US bought all gold offered at that price.

1939 London gold market closed on 3 September on outbreak of World War II. Final fix £8.05.

1944 Bretton Woods Agreement established new international framework of fixed exchange rates with gold exchanged for currencies among central banks at $35.

1949 US gold reserves peaked at 707 million troy/oz /21,990 m.t; equal to 75% of world stocks.

1954 London Gold Fixing resumed; opening price £12.42. Aim was to keep price equivalent to $35.

1961 Gold Pool of US and main European central banks set up to defend $35 price, by selling at fixing to contain it.

1965 Private buying exceeded mine supply, making Gold Pool net sellers.

1968 Collapse of Gold Pool and defence of $35 price, after devaluation of sterling and pressure on dollar over Vietnam setbacks sent speculators into gold. The pool lost almost 64 million troy/oz /2,000 m.t. London market closed for two weeks; when it reopened the fix was in dollars, not sterling, and an afternoon fix was added for New York's benefit. Gold price floated freely, but central banks still exchanged at $35.

1971 Federal Reserve in New York closed its 'gold window' at which central banks had still been able to trade dollars for gold at $35, ending the gold exchange standard.

1974 Hong Kong gold market liberalised. London market members soon opened trading rooms.

1975 American citizens again permitted to own gold after 42 years. Comex 1 kilo contract launched. US Treasury began five years of gold sales.

1976 IMF began four-year series of gold auctions.

1980 Record London fixing at $850 on 21 January ended an inflationary decade of oil price shocks, the freezing of Iran's assets and the Soviet invasion of Afghanistan, which sent investors into gold. Average price for the year was $614.63.

1987 The London Bullion Market Association founded to represent the interests of the members of the wholesale bullion market.

1999 The Euro was launched, with the European Central Bank holding 15% of its reserves in gold. The Bank of England announced the sale of half of the UK's gold stock. The Washington Agreement on Gold (The Central Bank Gold Agreement) set a five-year term of limited gold sales by central banks to stabilize the market.

2004 N. M. Rothschild & Sons gave up their seat at the fixing, which was taken by Barclays. Bank of Nova Scotia (Scotia Mocatta) became the new chair, with the fixing itself becoming a telephone process.

Appendix 3

U.S. Emergency War Powers Act 1933

1933 – Cuba. During a revolution against President Gerardo Machado naval

forces demonstrated but no landing was made.

1934 – China. Marines landed at Foochow to protect the American Consulate.

1940-1945- Newfoundland, Bermuda, St. Lucia, - Bahamas, Jamaica, Antigua, Trinidad, and British Guiana. Troops were sent to guard air and naval bases

obtained under lease by negotiation with the United Kingdom. These were sometimes called lend-lease bases but were under the Destroyers for Bases Agreement.

1941 – Greenland. Greenland was taken under protection of the United States in April.

1941 – Netherlands (Dutch Guiana). In November the President ordered American troops to occupy Dutch Guiana, but by agreement with the Netherlands government in exile, Brazil cooperated to protect aluminum ore supply from the bauxite mines in Suriname.

1941 – Iceland. Iceland was taken under the protection of the United States, with consent of its government replacing British troops, for strategic reasons.

1941 – Germany. Sometime in the spring the

President ordered the Navy to patrol ship lanes to Europe. By July US warships were convoying and by September were attacking German submarines. In November, the Neutrality Act was partly repealed to protect US military aid to Britain.

1944-46 Temporary reoccupation of the Philippines during WWII and in preparation for previously scheduled independence.

1945-49 Occupation of South Korea and defeat of a leftist insurgency.

1946 – Trieste (Italy). President Truman ordered the increase of US troops along the zonal occupation line and the reinforcement of air forces in northern Italy after Yugoslav forces shot down an unarmed US Army transport plane flying over Venezia Giulia. Earlier US naval units had been sent to the scene.Later the Free Territory of Trieste, Zone A.

1945-47 US Marines garrisoned in mainland China to oversee the removal of Soviet and Japanese forces after World War II.

1948 – Palestine. A marine consular guard was sent to Jerusalem to protect the US Consul General.

1948 – Berlin. Berlin Airlift After the Soviet Union established a land blockade of the US, British, and French sectors of Berlin on June 24, 1948, the United States and its allies airlifted supplies to Berlin until after the blockade was lifted in May 1949.

1950-1959– China. Marines were dispatched to Nanking to protect the American Embassy when the city fell to Communist troops, and to Shanghai to aid in the protection and evacuation of Americans.

1950-53 – Korean War

1950- The United States responded to North Korean invasion of South Korea by going to its assistance, pursuant to United Nations Security Council resolutions. US forces deployed in Korea exceeded 300,000 during the last year of the conflict. Over 36,600 US military were killed in action.

1950-55 – Formosa (Taiwan). In June 1950 at the beginning of the Korean War, President Truman ordered the US Seventh Fleet to prevent Chinese Communist attacks upon Formosa and Chinese Nationalist operations against mainland China.

1954-55 – China. Naval units evacuated US civilians and military personnel from the Tachen Islands.

1955-64 – Vietnam. First military advisors sent to Vietnam on 12 Feb 1955.

By 1964, US troop levels had grown to 21,000. On 7 August 1964, US Congress approved Gulf of Tonkin resolution affirming "All necessary measures to repel any armed attack against the forces of the United States. . .to prevent further aggression. . . (and) assist any member or protocol state of the Southeast Asian Collective Defense Treaty (SEATO) requesting assistance. . ."

1956 – Egypt. A marine battalion evacuated US nationals and other persons

from Alexandria during the Suez crisis.

1958 – Lebanon. Lebanon crisis of 1958 Marines were landed in Lebanon at
the invitation of President Camille Chamoun to help protect against threatened insurrection supported from the outside. The President's action was supported by a Congressional resolution passed in 1957 that authorized such actions in that area of the world.

1960-1969

1959-60 – The Caribbean. Second Marine Ground Task Force was deployed to protect US nationals following the Cuban revolution.

1941-45 – World War II

1941- the United States declared war with Japan in response to the bombing of Pearl Harbor. The US declared war against Bulgaria, Germany, Hungary, Italy and Romania in response to the declarations of war by those nations against the United States.

1945 – China. In October 50,000 US Marines were sent to North China to assist Chinese Nationalist authorities in disarming and repatriating the Japanese in China and in controlling ports, railroads, and airfields. This was in addition to approximately 60,000 US forces remaining in China at the end of

World War II 1945-1949

1945-49 Occupation of part of Germany.

1945-55 Occupation of part of Austria.

1945-46 Occupation of part of Italy.

1945-52 Occupation of Japan.

1962 – Thailand. The Third Marine Expeditionary Unit landed on May 17, 1962 to support that country during the threat of Communist pressure from outside; by July 30, the 5,000 marines had been withdrawn.

1962 – Cuba. Cuban Missile Crisis On October 22, President Kennedy instituted a "quarantine" on the shipment of offensive missiles to Cuba from the Soviet Union. He also warned Soviet Union that the launching of any missile from Cuba against nations in the Western Hemisphere would bring about USnuclear retaliation on the Soviet Union. A negotiated settlement was achieved in a few days.

1962-75 – Laos. From October 1962 until 1975, the United States played an important role in military support of anti-Communist forces in Laos.

1964 – Congo (Zaire). The United States sent four transport planes to provide airlift for Congolese troops during a rebellion and to transport Belgian paratroopers to rescue foreigners.

1959-75 – Vietnam War

1959-US military advisers had been in South Vietnam for a decade, and theirnumbers had been increased as the military position of the Saigon government became weaker. After citing what he termed were attacks on US destroyers in the Tonkin Gulf, President Johnson asked in August 1964 for a resolution expressing US determination to support freedom and protect peace in Southeast Asia. Congress responded with the Tonkin Gulf Resolution, expressing support for "all necessary measures" the President might take to repel armed attacks against US forces and prevent further aggression. Following this resolution, and following a Communist attack on a US installation in central Vietnam, the United States escalated its participation in the war to a peak of 543,000 military personnel by April 1969.

1965 – Dominican Republic. Invasion of Dominican Republic The United States intervened to protect lives and property during a Dominican revolt and sent 20,000 US troops as fears grew that the revolutionary forces were coming increasingly under Communist control.

1967 – Israel. The USS Liberty incident, whereupon a United States Navy Technical Research Ship was attacked June 8, 1967 by Israeli armed forces, killing 34 and wounding more than 170 U.S. crew members.

1967 – Congo (Zaire). The United States sent three military transport aircraft with crews to provide the Congo central government with logistical support during a revolt.[RL30172]

1968 – Laos & Cambodia. U.S. starts secret bombing campaign against targets along the Ho Chi Minh trail in the sovereign nations of Cambodia and Laos. The bombings last at least two years. (See Operation Commando Hunt)

1970-1979

1970 – Cambodia Campaign. US troops were ordered into Cambodia to clean out Communist sanctuaries from which Viet Cong and North Vietnamese attacked US and South Vietnamese forces in Vietnam. The object of this attack, which lasted from April 30 to June 30, was to ensure the continuing safe withdrawal of American forces from South Vietnam and to assist the program of Vietnamization.

1973 – Operation Nickel Grass, a strategic airlift operation conducted by the United States to deliver weapons and supplies to Israel during the Yom Kippur War.

1974 – Evacuation from Cyprus. United States naval forces evacuated US civilians during the Turkish invasion of Cyprus.

1975 – Evacuation from Vietnam. On April 3, 1975, President Ford reported US naval vessels, helicopters, and Marines had been sent to assist in evacuation of refugees and US nationals from Vietnam.

1975 – Evacuation from Cambodia. On April 12, 1975, President Ford reported that he had ordered US military forces to proceed with the planned evacuation of US citizens from Cambodia.

1975 – South Vietnam. On April 30, 1975, President Ford reported that a force
of 70 evacuation helicopters and 865 Marines had evacuated about 1,400 US citizens and 5,500 third country nationals and South Vietnamese from landing
zones in and around the US Embassy, Saigon and Tan Son Nhut Airport.

1975 – Cambodia. Mayagüez Incident. On May 15,

1975, President Ford reported he had ordered military forces to retake the SS Mayagüez, a merchant vessel which was seized from Cambodian naval patrol boats in international waters and forced to proceed to a nearby island.

1976 – Lebanon. On July 22 and 23, 1974, helicopters from five US naval vessels evacuated approximately 250 Americans and Europeans from Lebanon during fighting between Lebanese factions after an overland convoy evacuation had been blocked by hostilities.

1976 – Korea. Additional forces were sent to Korea after two American soldiers were killed by North Korean soldiers in the demilitarized zone between North and South Korea while cutting down a tree.

1978 – Zaire (Congo). From May 19 through June 1978, the United States utilized military transport aircraft to provide logistical support to Belgian and French rescue operations in Zaire.

1980-1990

1980 – Operation Eagle Claw, Iran. On April 26, 1980, President Carter reported the use of six U.S. transport planes and eight helicopters in an unsuccessful attempt to rescue the American hostages in Iran.

1981 – El Salvador. After a guerrilla offensive against the government of El Salvador, additional US military advisers were sent to El Salvador, bringing the total to approximately 55, to assist in training government forces in counterinsurgency.

1981 – Libya. First Gulf of Sidra Incident On August 19, 1981, US planes based on the carrier USS Nimitz shot down two Libyan jets over the Gulf of Sidra after one of the Libyan jets had fired a heat-seeking missile. The United States periodically held freedom of navigation exercises in the Gulf of Sidra, claimed by Libya as territorial waters but considered international waters by the United States.

1982 – Sinai. On March 19, 1982, President Reagan reported the deployment of military personnel and equipment to participate in the Multinational Force and Observers in the Sinai. Participation had been authorized by the Multinational Force and Observers Resolution, Public Law 97-132.

1982 – Lebanon. Multinational Force in Lebanon. On August 21, 1982, President Reagan reported the dispatch of 80 Marines to serve in the multinational force to assist in the withdrawal of members of the Palestine Liberation force from Beirut. The Marines left September 20, 1982.

1982-1983 – Lebanon. On September 29, 1982, President Reagan reported the deployment of 1200 marines to serve in a temporary multinational force to facilitate the restoration of Lebanese government sovereignty. On September 29, 1983, Congress passed the Multinational Force in Lebanon Resolution (P.L. 98-119) authorizing the continued participation for eighteen months.

1983 – Egypt. After a Libyan plane bombed a city

in Sudan on March 18, 1983, and Sudan and Egypt appealed for assistance, the United States dispatched an AWACS electronic surveillance plane to Egypt.

1983 – Grenada. Citing the increased threat of Soviet and Cuban influence and noting the development of an international airport following a bloodless Grenada coup d'état and alignment with the Soviets and Cuba, the U.S. launches Operation Urgent Fury to invade the sovereign island nation of Grenada.

1983-89 – Honduras. In July 1983 the United States undertook a series of exercises in Honduras that some believed might lead to conflict with Nicaragua. On March 25, 1986, unarmed US military helicopters and crewmen ferried Honduran troops to the Nicaraguan border to repel Nicaraguan troops.

1983 – Chad. On August 8, 1983, President Reagan reported the deployment of two AWACS electronic surveillance planes and eight F-15 fighter planes and ground logistical support forces to assist Chad against Libyan and rebel forces.

1984 – Persian Gulf. On June 5, 1984, Saudi Arabian jet fighter planes, aided by intelligence from a US AWACS electronic surveillance aircraft and fueled by a U.S. KC-10 tanker, shot down two Iranian fighter planes over an area of the Persian Gulf proclaimed as a protected zone for shipping.

1985 – Italy. On October 10, 1985, US Navy pilots intercepted an Egyptian airliner and forced it to land in Sicily. The airliner was carrying the hijackers of the

Italian cruise ship Achille Lauro who had killed an American citizen during the hijacking.

1986 – Libya. Action in the Gulf of Sidra (1986) On March 26, 1986, President Reagan reported on March 24 and 25, US forces, while engaged in freedom of navigation exercises around the Gulf of Sidra, had been attacked by Libyan missiles and the United States had responded with missiles.

1986 – Libya. Operation El Dorado Canyon On April 16, 1986, President Reagan reported that U.S. air and naval forces had conducted bombing strikes on terrorist facilities and military installations in the Libyan capitol of Tripoli, claiming that Libyan leader Col. Muammar al-Gaddafi was responsible for a bomb attack at a German disco that killed two U.S. soldiers.

1986 – Bolivia. U.S. Army personnel and aircraft assisted Bolivia in anti-drug operations.

1987 – Persian Gulf. USS Stark was struck on May 17 by two Exocet antiship missiles fired from an Iraqi F-1 Mirage during the Iran-Iraq War killing 37 US Navy sailors.

1987 –October 19, Operation Nimble Archer - attack on two Iranian oil platforms in the Persian Gulf by United States Navy forces. The attack was a response to Iran's October 16, 1987 attack on the MV Sea Isle City, a reflagged Kuwaiti oil tanker at anchor off Kuwait, with a Silkworm missile.

1987-88 – Persian Gulf. After the Iran-Iraq War resulted in several military incidents in the Persian Gulf, the United States increased US joint military forces operations in the Persian Gulf and adopted a policy of reflagging and escorting Kuwaiti oil tankers through the Persian Gulf, called Operation Earnest Will. President Reagan reported that US ships had been fired upon or struck mines or taken other military action on September 21 (Iran Ajr),

October 8, and October 19, 1987 and April 18 (Operation Praying Mantis),

July 3, and July 14, 1988. The United States gradually reduced its forces after a cease-fire between Iran and Iraq on August 20, 1988.[RL30172] It was the largest naval convoy operation since World War II.

1987-88 – Operation Earnest Will was the U.S. military protection of Kuwaiti oil tankers from Iraqi and Iranian attacks in 1987 and 1988 during the Tanker War phase of the Iran-Iraq War. It was the largest naval convoy operation since World War II.

1987-88 – Operation Prime Chance was a United States Special Operations Command operation intended to protect U.S. -flagged oil tankers from Iranian attack during the Iran-Iraq War. The operation took place roughly at the same time as Operation Earnest Will.

1988 – Operation Praying Mantis was the April 18, 1988 action waged by U.S. naval forces in retaliation for the Iranian mining of the Persian Gulf and the subsequent damage to an American warship.

1988 – Operation Golden Pheasant was an emergency deployment of U.S. troops to Honduras in

1988, as a result of threatening actions by the forces of the (then socialist) Nicaraguans.

1988 – USS Vincennes shoot down of Iran Air Flight 655 1988 – Panama. In mid-March and April 1988, during a period of instability in Panama and as the United States increased pressure on Panamanian head of state General Manuel Noriega to resign, the United States sent 1,000 troops to Panama, to "further safeguard the canal, US lives, property and interests in the area." The forces supplemented 10,000 US military personnel already in the Panama Canal Zone.

1989 – Libya. Second Gulf of Sidra Incident On January 4, 1989, two US Navy F-14 aircraft based on the USS John F. Kennedy shot down two Libyan jet fighters over the Mediterranean Sea about 70 miles north of Libya. The US pilots said the Libyan planes had demonstrated hostile intentions.

1989 – Panama. On May 11, 1989, in response to General Noriega's disregard of the results of the Panamanian election, President Bush ordered a brigade-sized force of approximately 1,900 troops to augment the estimated 11,000 U.S. forces already in the area.

1989 – Colombia, Bolivia, and Peru. Andean Initiative in War on Drugs. On September 15, 1989, President Bush announced that military and law enforcement assistance would be sent to help the Andean nations of Colombia, Bolivia, and Peru combat illicit drug producers and traffickers. By mid- September there were 50-100 US military advisers in Colombia in connection with transport and training in the use of military equipment, plus seven Special Forces teams of 2-12 persons to train troops in the three countries.[RL30172]

1989 – Operation Classic Resolve, Philippines - On December 2, 1989, President Bush reported that on December 1, Air Force fighters from Clark Air Base in Luzon had assisted the Aquino government to repel a coup attempt. In addition, 100 marines were sent from U.S. Naval Base Subic Bay to protect the United States Embassy in Manila.

1989-90 – Operation Just Cause, Panama - On December 21, 1989, President Bush reported that he had ordered US military forces to Panama to protect the lives of American citizens and bring General Noriega to justice. By February 13, 1990, all the invasion forces had been withdrawn.[RL30172] Around 200 Panamanian civilians were reported killed. The Panamanian head of state, General Manuel Noriega, was captured and brought to the U.S.

1990 – Liberia. On August 6, 1990, President Bush reported that a reinforced rifle company had been sent to provide additional security to the US Embassy in Monrovia, and that helicopter teams had evacuated U.S. citizens from Liberia.[RL30172]

1990 – Saudi Arabia. On August 9, 1990, President Bush reported that he had ordered the forward deployment of substantial elements of the US armed forcesinto the Persian Gulf region to help defend Saudi

Arabia after the August 2 invasion of Kuwait by Iraq. On November 16, 1990, he reported the continued buildup of the forces to ensure an adequate offensive military option. American hostages being held in Iran.

Recommended Reading